Disclaimer

The publisher of this book is by no way associated with the National Institute of Standards and Technology (NIST). The NIST did not publish this book. It was published by 50 page publications under the public domain license.

50 Page Publications.

Book Title: CFAST - Consolidated Model of Fire Growth and Smoke Transport (Version 6), Technical Reference Guide, APRIL 2009 REVISION

Book Author: Walter W. Jones; Richard D. Peacock; Glenn P. Forney; Paul A. Reneke

Book Abstract: CFAST is a two-zone fire model capable of predicting the environment in a multi-compartment structure subjected to a fire. It calculates the time evolving distribution of smoke and fire gases and the temperature throughout a building during a user-prescribed fire. This report describes the equations which constitute the model, the physical basis for these equations, and an evaluation of the sensitivity and predictive capability of the model. This report is an assessment of the model following the outline set forth in ASTM E1355, Standard Guide for Evaluating the Predictive Capability of Deterministic Fire Models.

Citation: NIST SP - 1026

Keyword: Fire growth, smoke transport, computer models, fire models, fire research, hazard assessment, toxicity

NIST Special Publication 1026

April 2009 Revision

CFAST – Consolidated Model of Fire Growth and Smoke Transport (Version 6)
Technical Reference Guide

Walter W. Jones
Richard D. Peacock
Glenn P. Forney
Paul A. Reneke

NIST National Institute of Standards and Technology • U.S. Department of Commerce

NIST Special Publication 1026
April 2009 Revision

CFAST – Consolidated Model of Fire Growth and Smoke Transport (Version 6)
Technical Reference Guide

Walter W. Jones
Richard D. Peacock
Glenn P. Forney
Paul A. Reneke
Fire Research Division
Building and Fire Research Laboratory

April 29, 2009
SVN Repository Revision: 125

U.S. Department of Commerce
Carlos M. Gutierrez, Secretary

National Institute of Standards and Technology
James M. Turner, Deputy Director

Disclaimer

The U. S. Department of Commerce makes no warranty, expressed or implied, to users of CFAST and associated computer programs, and accepts no responsibility for its use. Users of CFAST assume sole responsibility under Federal law for determining the appropriateness of its use in any particular application; for any conclusions drawn from the results of its use; and for any actions taken or not taken as a result of analyses performed using these tools. CFAST is intended for use only by those competent in the field of fire safety and is intended only to supplement the informed judgment of a qualified user. The software package is a computer model which may or may not have predictive value when applied to a specific set of factual circumstances. Lack of accurate predictions by the model could lead to erroneous conclusions with regard to fire safety. All results should be evaluated by an informed user.

Intent and Use

The algorithms, procedures, and computer programs described in this report constitute a methodology for predicting some of the consequences resulting from a prescribed fire. They have been compiled from the best knowledge and understanding currently available, but have important limitations that must be understood and considered by the user. The program is intended for use by persons competent in the field of fire safety and with some familiarity with personal computers. It is intended as an aid in the fire safety decision-making process.

Preface

CFAST is a two-zone fire model used to calculate the evolving distribution of smoke, fire gases and temperature throughout compartments of a constructed facility during a fire. In CFAST, each compartment is divided into two gas layers.

The modeling equations used in CFAST take the mathematical form of an initial value problem for a system of ordinary differential equations (ODEs). These equations are derived using the conservation of mass, the conservation of energy (equivalently the first law of thermodynamics), the ideal gas law and relations for density and internal energy. These equations predict as functions of time quantities such as pressure, layer height and temperatures given the accumulation of mass and enthalpy in the two layers. The CFAST model then consists of a set of ODEs to compute the environment in each compartment and a collection of algorithms to compute the mass and enthalpy source terms required by the ODEs.

In general, this document provides the technical documentation for CFAST along with significant information on validation of the model. It follows the ASTM E1355 guide for model assessment. The guide provides several areas of evaluation:

- Model and scenarios definition: CFAST is designed primarily to predict the environment within compartmented structures which results from unwanted fires. These can range from very small containment vessels, on the order of 1 m^3 to large spaces on the order of 1000 m^3. The appropriate size fire for a given application depends on the size of the compartment being modeled. A range of such validation exercises is discussed in chapter 6.

- Theoretical basis for the model: Details of the underlying theory, governing equations, correlations, and organization used in the model are presented. The process of development of the model is discussed with reference to a range of NIST memorandums, published reports, and peer-reviewed journal articles on the model. In addition to overall limitations of zone-fire modeling, limitations of the individual sub-models are discussed.

- Mathematical and numerical robustness: CFAST has been subjected to extensive use and review both internal to NIST and by users worldwide in a broad range of applications. In addition to review within NIST independent of the model developers, the model has been published in international peer-reviewed journals worldwide, and in industry-standard handbooks referenced in specific consensus standards. Besides formal internal and peer review, CFAST is subjected to continuous scrutiny because it is available to the general public and is used internationally by those involved in fire safety design and post-fire reconstruction.

- Model sensitivity: Many of the outputs from the CFAST model are relatively insensitive to uncertainty in the inputs for a broad range of scenarios. Not surprisingly, the heat release rate is the most important variable because it provides the driving force for fire driven flows. For CFAST, the heat release rate is prescribed by the user. Thus, careful selection of the fire size is necessary for accurate predictions. Other variables related to compartment geometry such as compartment height or vent sizes, while deemed important for the model outputs, are typically more easily defined for specific design scenarios than fire related inputs.

- Model evaluation: The CFAST model has been subjected to extensive validation studies by NIST and others. Although some differences between the model and the experiments were evident in these studies, they are typically explained by limitations of the model and uncertainty of the experiments. Most prominent in the studies reviewed was the overprediction of gas temperature often attributed to uncertainty in soot production and radiative fraction. Still, studies typically show predictions accurate within about 30% of measurements for a range of scenarios. Like all predictive models, the best predictions come with a clear understanding of the limitations of the model and of the inputs provided to the calculations.

Nomenclature

A	area, m^2
A_{slab}	cross-sectional area of vent slab in horizontal vent flow, m^2
A_v	cross-sectional area of a vent, m^2
C	vent constriction (or flow) coefficient, dimensionless
C_{LOL}	lower oxygen limit coefficient, fraction of the available fuel which can be burned with the available oxygen, dimensionless
C_T	constant from plume centerline temperature calculation, 9.115
c_p	heat capacity of air at constant pressure, J/kgK
c_v	heat capacity of air at constant volume, J/kgK
D	fire diameter, m
D^*	characteristic fire diameter parameter, $\left(Q_f / \rho_\infty c_p T_\infty \sqrt{g} \right)^{2/5}$
	vent diameter, m
d_0	inlet ceiling jet depth in corridor flow, m
E_O	energy release per unit mass of oxygen consumed, J/kg
E_i	internal energy in layer i, W
F_{k-j}	configuration factor, fraction of radiation given off by surface k intercepted by surface j, dimensionless
g	gravitational constant, 9.8 m/s^2
h	convective heat transfer coefficient (W/m^2 K)
$\dot{h}_i$	rate of addition of enthalpy into layer i, J/s
$\dot{h}_L$	rate of addition of enthalpy into lower layer in a compartment, J/s
$\dot{h}_U$	rate of addition of enthalpy into upper layer in a compartment, J/s
H	height of a compartment, m
	flame height, m
H_1	distance from fire source to target location in plume centerline temperature calculation, m
H_2	distance from virtual fire source to target location in plume centerline temperature calculation, m
H_c	heat of combustion of the fuel, J/kg
k	equivalent thermal conductivity of air, W/m K, with subscripts c, e and s
k	equilibrium coefficient for HCl transport and deposition
L	length of a compartment, m
	characteristic length for radiation calculation, m
m	mass, kg
$\dot{m}_e$	entrainment rate, kg/s

$\dot{m}_{ex}$	bi-directional vent flow in vertical flow vent, kg/s
$\dot{m}_f$	pyrolysis rate of the fire, kg/s
m_i	total mass in gas layer i, kg
$\dot{m}_{io}$	mass flow rate through a vent, kg/s
m_L	total mass in lower gas layer in a compartment, kg
$\dot{m}_O$	oxygen required for full combustion of available fuel, kg/s
$\dot{m}_p$	plume flow rate, kg/s
m_U	total mass in upper gas layer in a compartment, kg
P	pressure at floor level of a compartment, Pa
P_b	cross-vent differential pressure at the bottom of a vent flow slab in horizontal vent flow, Pa
P_t	cross-vent differential pressure at the top of a vent flow slab in horizontal vent flow, Pa
Q_{ceil}	average convective heat transfer from the ceiling jet to the ceiling surface, W
Q_f	total heat release rate of the fire, W
$Q_{f,C}$	heat release rate of the fire released as convective energy, W
$Q_{f,eq}$	effective heat release rate of a vent fire, kW
$Q_{f,R}$	heat release rate of the fire released as radiation, W
Q_{spray}	spread density of a sprinkler
$Q^*_{I,1}$	original fire strength for plume temperature calculation, dimensionless
$Q^*_{I,2}$	modified fire strength for plume temperature calculation when target location is in the upper layer, dimensionless
$\Delta \hat{q}_k$	net radiative flux at wall segment k
q	heat flux, W/m^2
R	universal gas conference, J/kgK
RTI	thermal characteristic response time index of a sprinkler or heat detector (m$^{1/2}$ s$^{1/2}$)
r	radial distance from the fire, m
S	vent coefficient for vertical flow vents, dimensionless
T_∞	ambient gas temperature in compartment well removed from a target, K
T_{amb}	ambient temperature, K
T_i	gas temperature of layer i, K
T_L	gas temperature of lower layer in a compartment, K
T_p	gas temperature in the plume, K
T_U	gas temperature of upper layer in a compartment, K
T_0	inlet gas temperature of the ceiling jet in corridor flow, K
T^*_1	calculated plume temperature at the transition between continuous flaming and intermittent flaming, K
T^*_2	calculated plume temperature at the transition intermittent flaming and the fire plume, K
t	time, s
V	total volume of a compartment, m^3
V_L	total volume of lower layer in a compartment, m^3
V_U	total volume of upper layer in a compartment, m^3
V_i	volume of gas layer i, m^3

v	velocity, m/s
v_0	inlet ceiling jet velocity in corridor flow, m/s
W	width of a compartment, m
W_f	molar mass of fuel, kg/mol
W_s	molar mass of soot, kg/mol
Y_{LOL}	mass fraction of oxygen below which combustion will no longer occur, dimensionless
Y_{O_2}	mass fraction of oxygen in a gas layer, dimensionless
y_s	soot yield, mass of soot produced by the fire per unit mass of fuel, kg/kg
$Z_{I,1}$	distance from fire source to the interface between upper and lower layers, m
$Z_{I,2}$	distance from virtual fire source to the interface between upper and lower layers, m
z	height above the base of the fire, m
z_v	height of the virtual origin of a vent fire, m
z_p	reduced height of the plume of a vent fire, m
z_0	height of the virtual origin of fire, m
z_1^*	height above the base of the fire at the transition between continuous flaming and intermittent flaming, m
z_2^*	height above the base of the fire at the transition intermittent flaming and the fire plume, m
α	gas absorptance, dimensionless
	thermal diffusivity in conduction (m^2/s)
β	experimentally-determined constant in plume centerline temperature calculation, $\beta^2 = 0.913$
γ	ratio of c_p/c_v, dimensionless
ΔT	temperature rise, K
ε	emissivity
ν	stoichiometric coefficients for combustion reaction, dimensionless
	kinematic viscosity, m^2/s
ρ	density, kg/m^3
ρ_∞	density of gas well removed from a target, kg/m^3
ρ_{cj}	density of the ceiling jet gas, kg/m^3
ρ_i	density of gas layer i, kg/m^3
σ	Stefan-Boltzman constant (5.67 x 10^{-8} W/m^{2}K^4)
τ	transmittance, dimensionless
χ^C	fraction of the fire's heat release rate released as convective energy, dimensionless
χ^R	fraction of the fire's heat release rate released as radiation, dimensionless
ξ	ratio of upper layer gas temperature to lower layer gas temperature, dimensionless

Acknowledgments

Continuing support for CFAST is via internal funding at NIST. In addition, support is provided by other agencies of the U.S. Federal Government, most notably the Nuclear Regulatory Commission Office of Research and the U.S. Department of Energy. The U.S. NRC Office of Research has funded key validation experiments, the preparation of the CFAST manuals, and the continuing development of sub-models that are of importance in the area of nuclear power plant safety. Special thanks to Mark Salley and Jason Dreisbach for their efforts and support. Support to refine the software development and quality assurance process for CFAST has been provided by the U.S. Department of Energy (DOE). The assistance of Subir Sen and Debra Sparkman in understanding DOE software quality assurance programs and the application of the process to CFAST is gratefully acknowledged. Thanks are also due to Allan Coutts, Washington Safety Management Solutions for his insight into the application of fire models to nuclear safety applications and detailed review of the CFAST document updates for DOE.

Contents

Disclaimer i

Intent and Use iii

Preface v

Nomenclature vii

Acknowledgments xi

1 Overview **1**
   1.1 History 1
   1.2 Model Evaluation 2

2 Model and Scenario Definition **5**
   2.1 Model Documentation 5
      2.1.1 Name and Version of the Model 5
      2.1.2 Type of Model 5
      2.1.3 Model Developers 6
      2.1.4 Relevant Publications 6
      2.1.5 Governing Equations and Assumptions 6
      2.1.6 Input Data Required to Run the Model 7
      2.1.7 Property Data 7
      2.1.8 Model Results 8
      2.1.9 Uses and Limitations of the Model 8
   2.2 Scenarios for which the Model is Evaluated in this Document 10
      2.2.1 Description of Scenarios of Interest 10
      2.2.2 List of Quantities Predicted by the Model 11
      2.2.3 Degree of Accuracy Required for Each Output Quantity 11

3 Theoretical Basis for CFAST **13**
   3.1 Derivation of Equations for a Two-Layer Model 14
   3.2 Equation Set Used in CFAST 16
   3.3 Limitations of the Zone Model Assumptions 17
   3.4 Source Terms for the Model 17
      3.4.1 The Fire 18

	3.4.2	Plumes	22
	3.4.3	Vent Flow	28
	3.4.4	Corridor Flow	35
	3.4.5	Heat Transfer	39
	3.4.6	Ceiling Jet	50
3.5	Heat Detection		52
3.6	Sprinkler Activation and Fire Attenuation		53
3.7	Species Concentration and Deposition		54
	3.7.1	Species Transport	54
	3.7.2	HCl Deposition	55
3.8	Single Zone Approximation		56
3.9	Review of the Theoretical Development of the Model		57
	3.9.1	Assessment of the Completeness of Documentation	58
	3.9.2	Assessment of Justification of Approaches and Assumptions	58
	3.9.3	Assessment of Constants and Default Values	59

4 Mathematical and Numerical Robustness — 61
- 4.1 Structure of the Numerical Routines — 61
- 4.2 Code Checking — 63
- 4.3 Numerical Tests — 63
- 4.4 Comparison with Analytic Solutions — 64

5 Sensitivity of the Model — 65
- 5.1 Factorial Design Studies — 65
  - 5.1.1 Model Inputs and Outputs — 66
  - 5.1.2 Sensitivity to Larger Changes in Model Inputs — 69
- 5.2 Response Surface Studies — 71
- 5.3 Latin Hypercube Sampling Studies — 74
- 5.4 Summary — 75

6 Survey of Past Validation Work — 77
- 6.1 Comparisons with Full-Scale Tests Conducted Specifically for the Chosen Evaluation — 77
- 6.2 Comparisons with Previously Published Test Data — 79
  - 6.2.1 NIST/NRC Verification and Validation — 80
  - 6.2.2 Fire Plumes — 80
  - 6.2.3 Multiple Compartments — 80
  - 6.2.4 Large Compartments — 81
  - 6.2.5 Prediction of Flashover — 81
- 6.3 Comparison with Documented Fire Experience — 83
- 6.4 Comparison with Experiments Which Cover Special Situations — 84
  - 6.4.1 Nuclear Facilities — 84
  - 6.4.2 Small Scale Testing — 85
  - 6.4.3 Unusual Geometry and Specific Algorithms — 85
- 6.5 Summary — 87

7	**Conclusion**	**89**
	References	**101**

List of Figures

1.1	Zone Model Terms.	2
3.1	Schematic of control volumes in a two-layer zone model.	14
3.2	Entrainment and burning in a two-layer, multi-compartment model.	19
3.3	Excess plume centerline temperature from Baum and McCaffrey correlation.	26
3.4	Geometry for plume centerline temperature calculation.	27
3.5	Vent opening size fraction as a function of time.	29
3.6	Geometry and notation for horizontal flow vents in a two-zone fire model.	30
3.7	Flow patterns and layer number conventions for horizontal flow through a vertical vent.	31
3.8	Some simple fan-duct systems.	34
3.9	Relative excess downstream temperature in a corridor using an adiabatic temperature boundary condition for several inlet depths and inlet temperature boundary conditions. The inlet velocity is given by eq(3.55).	38
3.10	Radiation Exchange in a two-zone fire model.	41
3.11	An example of the calculated two-wall (RAD2) and four-wall (RAD4) contributions to radiation exchange on a ceiling and wall surface.	42
3.12	Setup for a configuration factor calculation between two arbitrarily oriented finite areas.	43
3.13	Radiative heat transfer from a point source fire to a target.	45
3.14	Radiative heat transfer from the upper and lower layer gas to a target in the lower layer.	46
3.15	Convective heat transfer to ceiling and wall surfaces via the ceiling jet.	51
3.16	Schematic of hydrogen chloride deposition region.	55
4.1	Subroutine structure for the CFAST model.	62
5.1	Building Geometry for base case scenario.	68
5.2	An example of time dependent sensitivity of fire model outputs to a 10% change in room volume for a single room fire scenario.	69
5.3	Layer temperatures and volumes in several rooms resulting from variation in heat release rate for a four-room growing fire scenario.	70
5.4	Comparison of the time dependent heat release rate and layer temperatures in several rooms for a four-room growing fire scenario.	71
5.5	Sensitivity of temperature to heat release rate for a four-room growing fire scenario.	72
5.6	Sensitivity of temperature to heat release rate for a four-room growing fire scenario.	73

5.7 Sensitivity of temperature to heat release rate for a four-room growing fire scenario.   74

6.1 Comparison of correlations, CFAST predictions, and experimental data for the prediction of flashover in a compartment fire. 83

List of Tables

3.1	Conservative zone model equations.	16
3.2	Recommended compartment dimension limits.	17
3.3	Transfer coefficients for HCl deposition	56
5.1	Typical Inputs for a Two-Zone Fire Model	67
5.2	Typical Outputs for a Two-Zone Fire Model	67
6.1	Summary of Model Comparisons.	88

Chapter 1

Overview

1.1 History

Analytical models for predicting fire behavior have been evolving since the 1960s. Over the past decade, the completeness of the models has grown considerably. In the beginning, the focus of these efforts was to describe in mathematical language the various phenomena which were observed in fire growth and spread. These separate representations have typically described only a small part of a fire. However, when combined they can create a complex computational model that can predict the expected course of a fire.

Once a mathematical representation of the underlying physics has been developed, the conservation equations can be re-cast into predictive equations for temperature, smoke and gas concentration and other parameters of interest, and solved numerically. The equations are usually in the form of differential equations. A complete set of equations can describe the conditions produced by the fire at a given time in a specified volume of air. Referred to as a control volume, the model assumes that the predicted conditions within this volume are uniform at any time. Thus, the control volume has one temperature, smoke density, gas concentration, etc. Different models divide the building into different numbers of control volumes depending on the desired level of detail. The most common fire model, known as a zone model, generally uses two control volumes to describe a compartment — an upper layer and a lower layer. In the compartment with the fire, additional control volumes for the fire plume or the ceiling jet may be included to improve the accuracy of the prediction (see figure 1.1).

This two-layer approach has evolved from observation of such layering in real-scale fire experiments. Hot gases collect at the ceiling and fill the compartment from the top. While these experiments show some variation in conditions within the layer, these are small compared to the differences between the layers. Thus, the zone model can produce a fairly realistic simulation under many common and important conditions.

Other types of models include network models and field models. Network models use one control volume per compartment and are used to predict conditions in spaces far removed from the fire compartment where temperatures are near ambient and layering does not occur. The field model goes to the other extreme, dividing the compartment into thousands or millions of control volumes. Such models can predict the variation in conditions within the layers, but typically require far longer run times than zone models. They are used when a highly detailed prediction of

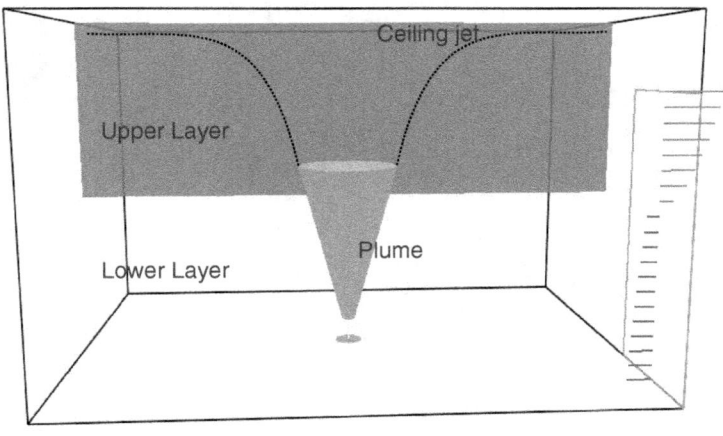

Figure 1.1: Zone Model Terms.

the flow itself is of interest.

1.2 Model Evaluation

The process of model evaluation is critical to establishing both the acceptable uses and limitations of fire models. It is not possible to evaluate a model in total; instead, available guides such as ASTM E1355 [1] are intended to provide a methodology for evaluating the predictive capabilities for a specific use. Validation for one application or scenario does not imply validation for different scenarios. Several alternatives are provided for performing the evaluation process including comparison of predictions against standard fire tests, full-scale fire experiments, field experience, published literature, or previously evaluated models.

The use of fire models currently extends beyond the fire research laboratory and into the engineering, fire service and legal communities. Sufficient evaluation of fire models is necessary to ensure that those using the models can judge the adequacy of the scientific and technical basis for the models, select models appropriate for a desired use, and understand the level of confidence which can be placed on the results predicted by the models. Adequate evaluation will help prevent the unintentional misuse of fire models. Verification is a process to check the correctness of the solution of the governing equations. Verification does not imply that the governing equations are appropriate; only that the equations are being implemented and solved correctly. Validation is a process to determine the appropriateness of the governing equations as a mathematical model of the physical phenomena of interest. Typically, validation involves comparing model results with experimental measurement. Differences that cannot be explained by numerical errors in the model or uncertainty in the experiments are attributed to the assumptions and simplifications of the physical model. These terms are used together to perform a model assessment. The more general term, model assessment, encompasses both verification and validation of a computer model.

In general, this document follows the ASTM E1355 [1] guide for model assessment and provides a model assessment for the zone fire model CFAST. The guide provides four areas of evaluation for predictive fire models:

- Defining the model and scenarios for which the evaluation is to be conducted (chapter 2),
- Assessing the appropriateness of the theoretical basis and assumptions used in the model (chapter 3),
- Assessing the mathematical and numerical robustness of the model (chapter 4), and
- Quantifying the uncertainty and accuracy of the model results in predicting the course of events in similar fire scenarios (chapters 5 and 6).

Chapter 2

Model and Scenario Definition

Sufficient documentation of calculation models is necessary to assess the adequacy of the scientific and technical basis of the model and the accuracy of the computational procedures for scenarios of interest. In addition, adequate documentation will help prevent the unintentional misuse of the model. The documentation in this document follows the guidelines in ASTM E1355-04 [1].

2.1 Model Documentation

2.1.1 Name and Version of the Model

The name of the model is the Consolidated Fire Growth and Smoke Transport Model or CFAST. The first public release was version 1.0 in June of 1990. This version was restructured from FAST [2] to incorporate the "lessons learned" from the zone model CCFM developed by Cooper and Forney [[3], namely that modification is easier and more robust if the components such as the physical routines are separated from the solver. chapter 4 (Mathematical and Numerical Robustness) discusses this in more detail. Version 2 was released as a component of Hazard 1.2 in 1994 [4,5]. The first of the 3.x series was released in 1995 and included a vertical flame spread algorithm, ceiling jets and nonuniform heat loss to the ceiling, spot targets, and heating and burning of multiple objects (ignition by flux, temperature or time) in addition to multiple prescribed fires. As it evolved over the next five years, version 3 included smoke and heat detectors, suppression through heat release reduction, better characterization of flow through doors and windows, vertical heat conduction through ceiling/floor boundaries, and non-rectangular compartments. In 2000, version 4 was released and included horizontal heat conduction through walls, and horizontal smoke flow in corridors. Version 5 improved the combustion chemistry. Version 6, released in July, 2005, incorporates a more consistent implementation of vents, fire objects and event processing and includes a graphical user interface which substantially improves its usability.

The code is written in FORTRAN 90.

2.1.2 Type of Model

CFAST is a model that predicts the environment within compartmented structures resulting from a fire prescribed by the user. It is an example of the class of models called finite element. This particular implementation is called a zone model, and essentially the space to be modeled is broken down to a few elements. The physics of the compartment fire phenomena is driven by fluid flow, primarily buoyancy. The usual set of elements or zones are the upper and lower gas layers, partitioning of the wall/ceiling/floor to an element each, one or more plumes and objects such as fires, targets, and detectors. One feature of this implementation of a finite element model is that the interface between the elements (in this case, the upper and lower gas layers) can move, with its position defined by the governing equations.

2.1.3 Model Developers

CFAST was developed and is maintained primarily by the Fire Research Division of the National Institute of Standards and Technology. The developers are Walter Jones, Richard Peacock, Glenn Forney, Rebecca Portier, Paul Reneke, and John Hoover [1].

There have been contributions through research and published papers at Worcester Polytechnic Institute, University of California at Berkeley, VTT of Finland and CITCM of France. An important guide to development of the model has been from many people around the world who have provided ideas, suggestions, comments, detailed questions, opinions on what should happen in particular scenarios, what physics and chemistry are needed and what types of problems must be addressed by such a model in order to be useful for real world applications.

2.1.4 Relevant Publications

To accompany the model and simplify its use, NIST has developed this Technical Reference Guide [6], a User's Guide [7] and a Software and Validation Guide [8]. The Technical Reference Guide describes the underlying physical principles and summarizes sensitivity analysis, model validation, and model limitations consistent with ASTM E1355 [1]. The Users Guide describes how to use the model.

The U.S. Nuclear Regulatory Commission has published a verification and validation study of five selected fire models commonly used in support of risk-informed and performance-based fire protection at nuclear power plants [9]. In addition to an extensive study of the CFAST model, the report compares the output of several other models ranging from simple hand calculations to more complex CFD codes such as the Fire Dynamics Simulator (FDS) developed by NIST.

There are documents available (http://cfast.nist.gov) that are applicable to versions 2, 3, 5 as well as 6 of both the model and user interface.

2.1.5 Governing Equations and Assumptions

For CFAST, as for most zone fire models, the equations solved are for conservation of mass and energy. The momentum equation is not solved explicitly, except for use of the Bernoulli equation for the flow velocity at vents. Based on an integration over the volume of an element, these equations are solved as ordinary differential equations.

There are two assumptions which reduce the computation time dramatically. The first is that relatively few zones or elements per compartment is sufficient to model the physical situation. The

---
[1] Naval Research Laboratory, Washington, DC 20375.

second assumption is to close the set of equations without using the momentum equation in the compartment interiors. This simplification eliminates acoustic waves. Though this prevents one from calculating gravity waves in compartments (or between compartments), coupled with only a few elements per compartment allows for a prediction in a large and complex space very quickly.

The equations themselves and the algorithms and sub-models used are discussed in detail in chapter 3.

2.1.6 Input Data Required to Run the Model

All of the data required to run the CFAST model reside in a primary data file, which the user creates. Some instances may require databases of information on objects, thermophysical properties of boundaries, and sample prescribed fire descriptions. In general, the data files contain the following information:

- compartment dimensions (height, width, length)
- construction materials of the compartment (e.g., concrete, gypsum)
- material properties (e.g., thermal conductivity, specific heat, density, thickness, heat of combustion)
- dimensions and positions of horizontal and vertical flow openings such as doors, windows, and vents
- mechanical ventilation specifications
- fire properties (e.g., heat release rate, lower oxygen limit, and species production rates as a function of time)
- sprinkler and detector specifications
- positions, sizes, and characteristics of targets

Sample data files are provided which encompass many of the validation exercises described in chapter 6 and in the various articles and reports referenced in chapter 6. These examples range from simple one-compartment simulations to a large multi-story hotel scenario that includes an elevator shaft and stairwell pressurization. A complete description of the input parameters required by CFAST can be found in the CFAST Users Guide [7]. Some of these parameters have default values included in the model, which are intended to be representative for a range of fire scenarios.

2.1.7 Property Data

Any simulation of a real fire scenario involves prescribing material properties for the walls, floor, ceiling, and furnishings. CFAST treats all of these materials as homogeneous solids, thus the physical parameters for many real objects can only be viewed as approximations to the actual properties. Describing these materials in the input data file is a challenging task for the model user. Thermal properties for the most common barrier materials used in construction, e.g. gypsum wallboard, are included in a database, thermal.df, included with the model. These properties come directly from handbook values for typical materials [10].

2.1.8 Model Results

The output of CFAST are the sensible variables that are needed for assessing the environment in a building subjected to a fire. Once the simulation is complete, CFAST produces an output file containing all of the solution variables. Typical outputs include (but are not limited to) the following:

- environmental conditions in the room (such as hot gas layer temperature; plume centerline temperature; oxygen and smoke concentration; and ceiling, wall, and floor temperatures)

- heat transfer-related outputs to walls and targets (such as incident convective, radiated, and total heat fluxes)

- fire intensity and flame height

- flow velocities through vents and openings

- detector and sprinkler activation times

There is more extensive discussion of the output in chapter 6 of this technical reference manual and the users guide. The output is always in the metric system of units.

2.1.9 Uses and Limitations of the Model

CFAST has been developed for use in solving practical fire problems in fire protection engineering. It is intended for use in system modeling of building and building components. A priori prediction of flame spread or fire growth on objects is not modeled. Rather, the consequences of a specified fire is estimated. It is not intended for detailed study of flow within a compartment, such as is needed for smoke detector siting. It includes the activation of sprinklers and fire suppression by water droplets.

The most extensive use of the model is in fire and smoke spread in complex buildings. The efficiency and computational speed are inherent in the few computation cells needed for a zone model implementation. The use is for design and reconstruction of time-lines for fire and smoke spread in residential, commercial, and industrial fire applications. Some applications of the model have been for design of smoke control systems.

- Compartments: CFAST is generally limited to situations where the compartment volumes are strongly stratified. However, in order to facilitate the use of the model for preliminary estimates when a more sophisticated calculation is ultimately needed, there are algorithms for corridor flow, smoke detector activation, and detailed heat conduction through solid boundaries. This model does provide for non-rectangular compartments, although the application is intended to be limited to relatively simple spaces. There is no intent to include complex geometries where a complex flow field is a driving force. For these applications, computational fluid dynamics (CFD) models are appropriate.

- Gas Layers: There are also limitations inherent in the assumption of stratification of the gas layers. The zone model concept, by definition, implies a sharp boundary between the upper and lower layers, whereas in reality, the transition is typically over about 10% of the height of the compartment and can be larger in weakly stratified flow. For example, a burning cigarette in a normal room is not within the purview of a zone model. While it is possible to make predictions within 5% of the actual temperatures of the gas layers, this is not the optimum use of the model. It is more properly used to make estimates of fire spread (not flame spread), smoke detection and contamination, and life safety calculations.

- Heat Release Rate: CFAST does not predict fire growth on burning objects. Heat release rate is specified by the user for one or more fire objects. The model does include the ability to limit the specified burning based on available oxygen. There are also limitations inherent in the assumptions used in application of the empirical models. As a general guideline, the heat release should not exceed about 1 MW/m^3. This is a limitation on the numerical routines attributable to the coupling between gas flow and heat transfer through boundaries (conduction, convection, and radiation). The inherent two-layer assumption is likely to break down well before this limit is reached.

- Radiation: Because the model includes a sophisticated radiation model and ventilation algorithms, it has further use for studying building contamination through the ventilation system, as well as the stack effect and the effect of wind on air circulation in buildings. Radiation from fires is modeled with a simple point source approximation. This limits the accuracy of the model near fire sources. Calculation of radiative exchange between compartments is not modeled.

- Ventilation and Leakage: In a single compartment, the ratio of the area of vents connecting one compartment to another to the volume of the compartments should not exceed roughly 2 m. This is a limitation on the plug flow assumption for vents. An important limitation arises from the uncertainty in the scenario specification. For example, leakage in buildings is significant, and this affects flow calculations especially when wind is present and for tall buildings. These effects can overwhelm limitations on accuracy of the implementation of the model. The overall accuracy of the model is closely tied to the specificity, care, and completeness with which the data are provided.

- Thermal Properties: The accuracy of the model predictions is limited by how well the user can specify the thermophysical properties. For example, the fraction of fuel which ends up as soot has an important effect on the radiation absorption of the gas layer and, therefore, the relative convective versus radiative heating of the layers and walls, which in turn affects the buoyancy and flow. There is a higher level of uncertainty of the predictions if the properties of real materials and real fuels are unknown or difficult to obtain, or the physical processes of combustion, radiation, and heat transfer are more complicated than their mathematical representations in CFAST.

User feedback indicates that using CFAST to predict the transport of heat and combustion products from a prescribed fire is straightforward, easily and quickly accomplished, and the results are within expectations. Any user of a computer based (numerical) model must be aware of the assumptions and approximations being employed. Except for those few materials supplied in the property databases, the user must supply the thermal properties of the materials, and then assess the performance of the model compared with experiments to ensure that the model is valid for a specific application. Only then can the model be expected to predict the outcome of fire scenarios that are similar to those that have actually been tested.

In addition, there are specific limitations and assumptions made in the development of the algorithms. These are detailed in the discussion of each of these sub-models:

In addition, there are specific limitations and assumptions made in the development of the algorithms. These are detailed in the discussion of each of these sub-models:

- section 3.3 on zone model assumptions,
- section 3.4.1 on prescribed fires,
- section 3.4.1 on the relationship between fires and mass balance,
- section 3.4.2 on the plume entrainment model,
- section 3.4.4 on the assumptions made for corridor flow correlations,
- section 3.4.5 on the assumptions made for radiation heat transfer,
- section 3.6 on the suppression model, and
- section 3.7.2 on HCl deposition.

2.2 Scenarios for which the Model is Evaluated in this Document

CFAST is used for a wide range of buildings of interest, from glove-box size compartments, to complex hotels to the vehicle assembly building at Cape Canaveral. The intended use of ASTM E1355 [1] is to validate a specific scenario of interest so that the model can be used for scenarios similar to the chosen scenario. The intent of this document, however, is to cover a much wider range of scenarios which encompass the range of acceptable use of the model. Thus, this section provides a description of this broader range of scenarios as discussed in this technical reference guide rather than a single, specific scenario of interest for a validation exercise.

2.2.1 Description of Scenarios of Interest

CFAST is designed primarily to predict the environment within compartmented structures which results from unwanted fires. These can range from very small containment vessels, on the order of $1 m^3$ to large spaces on the order of $1000 m^3$. As discussed in the section on limitations and use (see section 2.1.9), the appropriate size fire depends on the size of the compartment being modeled. A range of such validation exercises is discussed in chapter 6.

2.2.2 List of Quantities Predicted by the Model

CFAST provides a prediction of the plume centerline, gas layer, and boundary temperatures, target temperatures, species concentration (including soot volume fraction), layer height, fire size and flame length, floor pressure, flow and fire size at vents, and heat flux (both radiative and convective). There is a more extensive discussion of the output in the CFAST users guide.

2.2.3 Degree of Accuracy Required for Each Output Quantity

The degree of accuracy for each output variable required by the user is highly dependent on the technical issues associated with the analysis. The user must task: How accurate does the analysis have to be to answer the technical question posed? Thus, a generalized definition of the accuracy required for each quantity with no regard as to the specifics of a particular analysis is not practical and would be limited in its usefulness.

Returning to the earlier definitions of "design" and "reconstruction," fire scenarios, design applications typically are more accurate because the heat release rate is prescribed rather than predicted, and the initial and boundary conditions are far better characterized. Mathematically, a design calculation is an example of a "well-posed" problem in which the solution of the governing equations is advanced in time starting from a known set of initial conditions and constrained by a known set of boundary conditions. The accuracy of the results is a function of the fidelity of the numerical solution, which is largely dependent on the quality of the model inputs.

A reconstruction is an example of an "ill-posed" problem because the outcome is known whereas the initial and boundary conditions are not. There is no single, unique solution to the problem. Rather, it is possible to simulate numerous fires that produce the given outcome. There is no right or wrong answer, but rather a small set of plausible fire scenarios that are consistent with the collected evidence and physical laws incorporated into the model. These simulations are then used to demonstrate why the fire behaved as it did based on the current understanding of fire physics incorporated in the model. Most often, the result of the analysis is only qualitative. If there is any quantification at all, it could be in the time to reach critical events, like a roof collapse or room flashover.

The CFAST validation guide [8] includes efforts to date involving well-characterized geometries and prescribed fires. These studies show that CFAST predictions vary from being within experimental uncertainty to being about 30 % different than measurements of temperature, heat flux, gas concentration, *etc* (see, for example, reference [9]). In general, this is adequate for its intended uses which are life-safety calculations and estimation of the environment to which building elements are subjected in a fire environment. Applied design margins are typically larger than this level of accuracy and may be appropriate to insure an adequate factor of safety.

Chapter 3

Theoretical Basis for CFAST

Adequately detailed documentation of the theoretical basis of the model allows the model user to understand the underlying theory behind the model implementation and thus be able to assess the appropriateness of the model to specific problems. This chapter presents a derivation of the predictive equations for zone fire models and explains in detail the ones used in CFAST.

The modeling equations used in CFAST take the mathematical form of an initial value problem for a system of ordinary differential equations. These equations are derived using the conservation of mass, the conservation of energy (equivalently the first law of thermodynamics), the ideal gas law. These equations predict as functions of time quantities such as pressure, layer height and temperatures given the accumulation of mass and enthalpy in the two layers. The assumption of a zone model is that properties such as temperature can be approximated throughout a control volume by an average value.

Many formulations based upon these assumptions can be derived. One formulation can be converted into another using the definitions of density, internal energy and the ideal gas law. Though equivalent analytically, these formulations differ in their numerical properties. Each formulation can be expressed in terms of mass and enthalpy flow. These rates represent the exchange of mass and enthalpy between zones due to physical phenomena such as plumes, natural and forced ventilation, convective and radiative heat transfer, and so on. For example, a vent exchanges mass and enthalpy between zones in connected rooms, a fire plume typically adds heat to the upper layer and transfers entrained mass and enthalpy from the lower to the upper layer, and convection transfers enthalpy from the gas layers to the surrounding walls.

As discussed in references [11] and [12], the zone fire modeling ordinary differential equations (ODEs) are stiff. The term stiff means that large variations in time scales are present in the ODE solution. In our problem, pressures adjust to changing conditions more quickly than other to solve zone fire modeling ODEs because of this stiffness. Runge-Kutta methods or predictor-corrector methods such as Adams-Bashforth require prohibitively small timesteps in order to track the short-timescale phenomena (pressure in our case). Methods that calculate the Jacobian (or at least approximate it) have a much larger stability region for stiff problems and are thus more successful at their solution.

3.1 Derivation of Equations for a Two-Layer Model

A compartment is divided into two control volumes, a relatively hot upper layer and a relatively cool lower layer, as illustrated in figure 3.1. The gas in each layer has attributes of mass, internal energy, density, temperature, and volume denoted respectively by m_i, E_i, ρ_i, T_i, and V_i where $i=L$ for the lower layer and $i=U$ for the upper layer. The compartment as a whole has the attribute of pressure P. These 11 variables are related by means of the following seven constraints (counting density, internal energy and the ideal gas law twice, once for each layer).

$$\rho_i = \frac{m_i}{V_i} \qquad \text{Density} \qquad (3.1)$$

$$E_i = c_v m_i T_i \qquad \text{Internal Energy} \qquad (3.2)$$

$$P = R \rho_i T_i \qquad \text{Ideal Gas Law} \qquad (3.3)$$

$$V = V_L + V_U \qquad \text{Total Volume} \qquad (3.4)$$

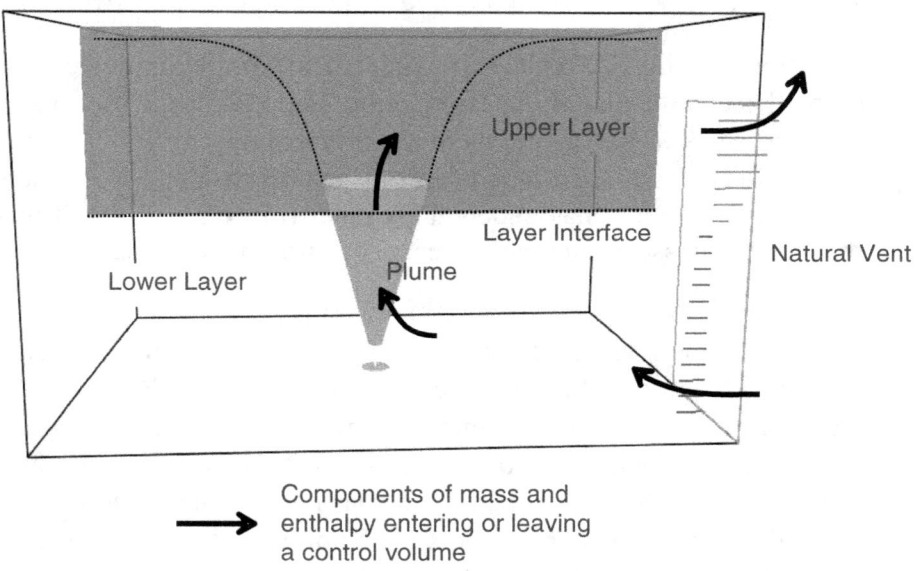

Figure 3.1: Schematic of control volumes in a two-layer zone model.

The specific heat at constant volume and at constant pressure c_v and c_p, the universal gas constant, R, and the ratio of specific heats, γ, are related by $\gamma = c_p/c_v$ and $R = c_p - c_v$. For ambient air, $c_p \approx 1 \text{kJ/kgK}$ and $\gamma = 1.4$. Four additional equations obtained from conservation of mass and energy for each layer are required to complete the equation set. The differential equations for mass in each layer are

$$\frac{dm_L}{dt} = \dot{m}_L \qquad (3.5)$$

$$\frac{dm_U}{dt} = \dot{m}_U \qquad (3.6)$$

The first law of thermodynamics states that the rate of increase of internal energy plus the rate at which the layer does work by expansion is equal to the rate at which enthalpy is added to the gas. In differential form this is

$$\underbrace{\frac{dE_i}{dt}}_{\text{internal energy}} + \underbrace{P\frac{dV_i}{dt}}_{\text{work}} = \underbrace{\dot{h}_i}_{\text{enthalpy}} \qquad (3.7)$$

where c_p is taken as constant in the enthalpy term,

$$\dot{h} = c_p \dot{m}_U T_U + \dot{E}_U + c_p \dot{m}_L T_L + \dot{E}_L \qquad (3.8)$$

A differential equation for pressure can be derived by adding the upper and lower layer versions of eq (3.7), noting that $\frac{dV_U}{dt} = -\frac{dV_L}{dt}$, and that

$$\frac{dE_i}{dt} = \frac{d(c_v \dot{m}_i T_i)}{dt} = \frac{c_v}{R}\frac{d(PV_i)}{dt} \qquad (3.9)$$

to obtain

$$\frac{dP}{dt} = \frac{\gamma - 1}{V}\left[\dot{h}_L + \dot{h}_U\right] \qquad (3.10)$$

Differential equations for the layer volumes can be obtained by substituting equation 3.9 into 3.7 to obtain

$$\frac{dV_i}{dt} = \frac{1}{P\gamma}\left[(\gamma - 1)\dot{h}_i - V_i \frac{dP}{dt}\right] \qquad (3.11)$$

Equation 3.2 can be rewritten using eq 3.11 to eliminate dV/dt to yield

$$\frac{dE_i}{dt} = \frac{1}{\gamma}\dot{h}_i + V\frac{dP}{dt} \qquad (3.12)$$

A differential equation for density can be derived by applying the quotient rule to $\frac{d\rho_i}{dt} = \frac{d}{dt}\frac{m_i}{V_i}$ and using eq 3.11 to eliminate $\frac{dV_i}{dt}$ to obtain

$$\frac{d\rho_i}{dt} = \frac{-1}{c_p T_i V_i}\left[\dot{h}_i - c_p \dot{m}_i T_i - \frac{V_i}{\gamma - 1}\frac{dP}{dt}\right] \qquad (3.13)$$

Temperature differential equations can be obtained from the equation of state by applying the quotient rule to $\frac{dT_i}{dt} = \frac{d}{dt}\frac{P}{R\rho_i}$ and using eq 3.13 to eliminate $\frac{d\rho_i}{dt}$ to obtain

$$\frac{dT_i}{dt} = \frac{1}{c_p \rho_i V_i}\left[\dot{h}_i - c_p \dot{m}_i T_i + V_i \frac{dP}{dt}\right] \qquad (3.14)$$

These equations for each of the 11 variables are summarized in table 3.1. The time evolution of these solution variables can be computed by solving the corresponding differential equations

Table 3.1: Conservative zone model equations

Equation Type	Differential Equation
i'th layer mass	$\dfrac{dm_i}{dt} = \dot{m}_i$
pressure	$\dfrac{dP}{dt} = \dfrac{\gamma - 1}{V} \dot{h}_L + \dot{h}_U$
i'th layer energy	$\dfrac{dE_i}{dt} = \dfrac{1}{\gamma} \dot{h}_i + V \dfrac{dP}{dt}$
i'th layer volume	$\dfrac{dV_i}{dt} = \dfrac{1}{P\gamma} \left((\gamma - 1)\dot{h}_i - V_i \dfrac{dP}{dt} \right)$
i'th layer density	$\dfrac{d\rho_i}{dt} = \dfrac{-1}{c_p T_i V_i} \left(\dot{h}_i - c_p \dot{m}_i T_i - \dfrac{V_i}{\gamma - 1} \dfrac{dP}{dt} \right)$
i'th layer temperature	$\dfrac{dT_i}{dt} = \dfrac{1}{c_p rho_i V_i} \left(\dot{h}_i - c_p \dot{m}_i T_i + V_i \dfrac{dP}{dt} \right)$

together with appropriate initial conditions. The remaining seven variables can be determined from the four solution variables using eqs (3.1) to (3.4).

There are, however, many possible differential equation formulations. Indeed, there are 330 different ways to select four variables from eleven. Many of these systems are incomplete due to the relationships that exist between the variables given in eqs (3.1) to (3.4). For example the variables, ρ_U, V_U, m_U, and P form a dependent set since $\rho_U = m_U / V_U$.

The number of differential equation formulations can be considerably reduced by not mixing variable types between layers; that is, if upper layer mass is chosen as a solution variable, then lower layer mass must also be chosen. For example, for two of the solution variables choose m_L and m_U, or ρ_L and ρ_U, or T_L and T_U. For the other two solution variables pick E_L and E_U or P and V_L or P and V_U. This reduces the number of distinct formulations to nine. Since the numerical properties of the upper layer volume equation are the same as a lower layer one, the number of distinct formulations can be reduced to six.

3.2 Equation Set Used in CFAST

The current version of CFAST is set up to use the equation set for layer temperature, layer volume, and pressure as shown below.

$$\frac{dP}{dt} = \frac{\gamma - 1}{V} \dot{h}_L + \dot{h}_U \tag{3.15}$$

$$\frac{dV_U}{dt} = \frac{1}{P\gamma} \left((\gamma - 1)\dot{h}_i - V_U \frac{dP}{dt} \right) \tag{3.16}$$

$$\frac{dT_U}{dt} = \frac{1}{c_p rho_i V_U} \left(\dot{h}_U - c_p \dot{m}_U T_U + V_U \frac{dP}{dt} \right) \tag{3.17}$$

$$\frac{dT_L}{dt} = \frac{1}{c_p rho_i V_L}\left[\dot{h}_L - c_p \dot{m}_L T_L\right] + V_L \frac{dP}{dt} \qquad (3.18)$$

In these equations, the pressure is actually modeled with the pressure difference relative to an ambient reference pressure to minimize numerical instability.

3.3 Limitations of the Zone Model Assumptions

The basic assumption of all zone fire models is that each compartment can be divided into a small number of control volumes, each of which is uniform in temperature and composition. In CFAST all compartments have two zones except for the fire room which has an additional zone for the plume. Since a real upper/lower interface is not as sharp as this, one has a spatial error of about 10 % in determining the height of the layer [13, 14].

The zone model concept best applies for an enclosure in which the width and length are not too different. If the horizontal dimensions of the room differ too much (i.e., the room looks like a corridor), the flow pattern in the room may become asymmetrical. If the enclosure is too shallow, the temperature may have significant radial differences. The width of the plume may at some height become equal to the width of the room and the model assumptions may fail in a tall and narrow enclosure. Therefore, the user should recognize approximate limits on the ratio of the length (L), width (W), and height (H) of the compartment.

If the aspect ratio (length/width) is over about 10, the corridor flow algorithm should be used. This provides the appropriate filling time. Similarly, for all shafts (elevators and stairways), a single zone approximation is more appropriate. It was found experimentally [15] that the mixing between a plume and lower layer due to the interaction with the walls of the shaft, caused complete mixing. The is the flipside of the corridor problem and occurs at a ratio of the height to characteristic floor length of about 10. The following quantitative limits are recommended:

Table 3.2: Recommended compartment dimension limits

Group	Acceptable	Special consideration required	Corridor flow algorithm
$(L/W)_{max}$	$L/W < 3$	$3 < L/W < 5$	$L/W > 5$
$(L/H)_{max}$	$L/H < 3$	$3 < L/H < 6$	$L/H > 6$
$(W/H)_{max}$	$W/H > 0.4$	$0.2 < L/W < 0.4$	$L/W < 0.2$

3.4 Source Terms for the Model

This section discusses each of the sub-models in CFAST. In general, the sections are similar to the way the model itself is structured. The sub-sections which follow discuss the way the actual phenomena are implemented numerically. For each of the phenomena discussed below, the physical basis for the model is discussed first, followed by a brief presentation of the implementation within

CFAST. For all of the phenomena, there are basically two parts to the implementation: the physical interface routine (which is the interface between the CFAST model and the algorithm) and the actual physical routine(s) which implement the physics. This implementation allows the physics to remain independent of the structure of CFAST and allows easier insertion of new phenomena.

3.4.1 The Fire

A fire in CFAST is implemented as a source of fuel mass which is released at a prescribed rate (the pyrolysis rate). Energy is released by the fuel and combustion products are created as it burns.

The model can simulate multiple fires in one or more compartments of the building. These fires are treated as totally separate entities, with no interaction of the plumes. These fires are generally referred to as "objects" and can be ignited at a prescribed time, temperature or heat flux.

CFAST does not include a pyrolysis model to predict fire growth. Rather, pyrolysis rates for each fire are prescribed by the user. While this approach does not directly account for increased pyrolysis due to radiative feedback from the flame or compartment, in theory these effects could be prescribed by the user. In an actual fire, this is an important consideration, and the specification used should consider the experimental conditions as closely as possible.

Constrained Fire

A fire releases energy based on the pyrolysis of fuel, but may be constrained by the oxygen available for combustion depending on the compartment conditions. Complete burning will take place only where there is sufficient oxygen. When insufficient oxygen is entrained into the fire plume, unburned fuel will be transported from zone to zone until there is sufficient oxygen and a high enough temperature to support combustion. In general, CFAST uses a simple definition of a combustion reaction that includes major products of combustion for hydrocarbon fuels:

$$C_xH_yO_zN_aCl_b + \nu_{O_2} O_2 \rightarrow \nu_{CO_2} CO_2 + \nu_{H_2O} H_2O + \nu_{CO} CO + \nu_s \text{Soot} + \nu_{N_2} N_2 + \nu_{HCl} HCl + \nu_{HCN} HCN \tag{3.19}$$

where the stoichiometric coefficients ν_{O_2}, ν_{CO_2}, etc. represent appropriate molar ratios for a stoichiometric balance of the equation. For example, for soot, it is related to the *soot yield*, y_s, via the relation:

$$\nu_s = \frac{W_F}{W_S} y_s \tag{3.20}$$

For complete combustion of the simplest hydrocarbon fuel, methane reacts with oxygen to form carbon dioxide and water. The only input required is the pyrolysis rate and the heat of combustion. For fuels that contain oxygen, nitrogen, or chlorine, the reaction becomes more complex. In this case, production yields for the species are prescribed by the user. Stoichiometry is used to insure conservation of mass and elements in the reaction. The species which are calculated are oxygen, carbon dioxide, carbon monoxide, water, and soot. Gaseous nitrogen is included, but only acts as a diluent. Production of hydrogen cyanide and hydrogen chloride are tracked solely based on user prescribed yields. There is a separate calculation of the concentration-time product Ct. Finally, a user-specified trace species can be specified to follow the transport that results from fire-induced flow for an arbitrary species. This may be of particular interest for radiological releases [16], but may be useful for any trace amounts released by a fire.

The heat release rate for a constrained fire may be reduced below its prescribed value based upon the oxygen available for combustion. When there is not enough oxygen to support complete combustion, some of the fuel will be transported to the gas layers and through vents as unburned hydrocarbons.

As fuel and oxygen are consumed, heat is released and various products of combustion are formed. The heat is released as radiation and convected enthalpy:

$$Q_{f,R} = \chi_R Q_f \qquad (3.21)$$
$$Q_{f,C} = (1 - \chi_R) Q_f \qquad (3.22)$$

where, χ_R is the fraction of the fires heat release rate given off as radiation. The convective enthalpy, $Q_{f,C}$ then becomes the driving term in the plume flow. For a constrained fire there is radiation to both the upper and lower layers, whereas the convective part contributes only to the upper layer.

Limiting Combustion by Available Oxygen

For any individual fire, the heat release rate is limited by available oxygen in the layer where the fire is located. This limit is applied in three places, which are shown schematically in figure 3. The first is burning in the portion of the plume which (at least initially) is typically in the lower layer of the room of fire origin (region #1). The second is the portion of the plume in the upper layer, also in the room of origin (region #2). The third is in the vent flow which entrains air from a lower layer into an upper layer in an adjacent compartment (region #3). The unburned hydrocarbons are tracked in this model. Further combustion of CO to CO_2 is not included in the model.

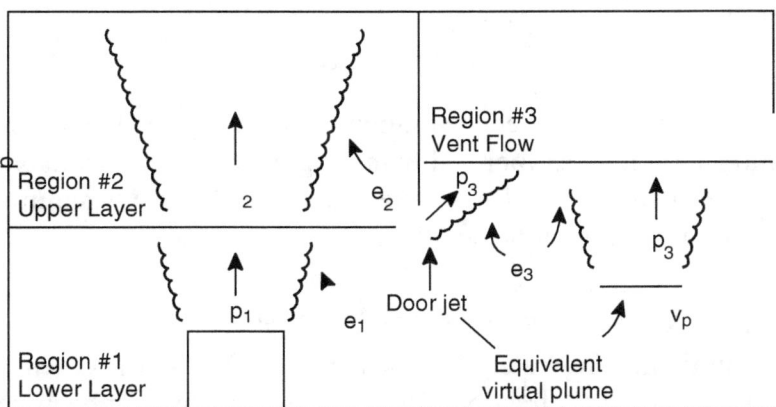

Figure 3.2: Entrainment and burning in a two-layer, multi-compartment model.

Initially, $\dot{m}_f$ is just the pyrolysis rate of the source fire in kg/s (region #1). For subsequent regions, the burning rate $\dot{m}_f$ is the unburned fuel from a previous region, $\dot{m}_{tuhc} = \dot{m}_f - \dot{m}_b$ where the subscript $tuhc$ is total unburned hydrocarbons, f is the fire source, and b is the amount burned.

The first step is to limit the actual burning which takes place in the combustion zone. In each combustion zone, there is a quantity of fuel available. At the source this results from the pyrolysis of the material, $\dot{m}_f$. In other situations such as a plume or door jet, it is the net unburned fuel available, $\dot{m}_{tuhc}$. In each case, the fuel which is available but not burned is then deposited into the "$\dot{m}_{tuhc}$" category. This provides a consistent notation. In the discussion below, $\dot{m}_f$ is the amount of fuel burned. This value is initially specified as to the available fuel, and then reduced if there is insufficient oxygen to support complete combustion. Subsequently, the available fuel, $\dot{m}_{tuhc}$, is reduced by the final value of $\dot{m}_f$ burned or $\dot{m}_b$. Thus we have a consistent description in each burning region, with an algorithm that is invoked independent of the region being analyzed.

$$Q_f = \dot{m}_f H_c \tag{3.23}$$

with the mass of oxygen required to achieve this energy release rate of

$$\dot{m}_O = \frac{Q_f}{E_O} = \dot{m}_f \frac{H_c}{E} \tag{3.24}$$

where E_O is the heat release per mass unit of oxygen consumed, taken to be 1.31×10^7 J/kg[1] (based on oxygen consumption calorimetry for typical fuels [17, 18, 19]). If the fuel contains oxygen (available for combustion), the oxygen needed to achieve full combustion is less:

$$\dot{m}_{O,needed} = \dot{m}_O - \dot{m}_{O,inthefuel} \tag{3.25}$$

If sufficient oxygen is available, then it is fully burned. However, if the oxygen concentration is low enough, it will constrain the burning and impose a limit on the amount of fuel actually burned, as opposed to the amount pyrolyzed. The actual limitation is discussed below and is:

$$\dot{m}_{O.actual} = \min{\dot{m}_{O,available}, \dot{m}_{O,needed}} \tag{3.26}$$

$$\dot{m}_{f,actual} = \dot{m}_{O,actual} \frac{E_O}{H_c} \tag{3.27}$$

The relationship between oxygen and fuel concentration defines a range in which burning will take place. In the CFAST model, a limit is incorporated by limiting the burning rate as the oxygen level decreases until a "lower oxygen limit" (LOL) is reached. The lower oxygen limit is incorporated through a smooth decrease in the burning rate near the limit:

$$\dot{m}_{O,available} = \dot{m}_e Y_{O_2} C_{LOL} \tag{3.28}$$

where $\dot{m}_e$ is the mass entrainment flow rate, Y_{O_2} is the mass fraction of oxygen, and the lower oxygen limit coefficient, C_{LOL}, is the fraction of the available fuel which can be burned with the available oxygen and varies from 0 at the limit to 1 above the limit. The functional form that utilizes the hyperbolic tangent was determined empirically to provide a smooth cutoff of the burning over a narrow range above the limit.

$$C_{LOL} = \frac{\tanh(800(Y_{O_2} - Y_{LOL}) - 400) + 1}{2} \tag{3.29}$$

---

[1] The units for oxygen consumption calorimetry are J/kg. The value 1.31×10^7 J/kg is representative of typical fuels such as furniture (see reference [17]) and implies these units. The variation or uncertainty (2σ) associated with this value is on the order of $\pm 5\%$

A temperature criterion is also imposed so that no burning will take place when the temperature is below a user prescribed temperature.

In summary, it is possible to follow the formation of the major products of combustion (carbon dioxide, carbon monoxide, soot, water, hydrogen cyanide, and hydrogen chloride) using appropriate measured product yields (e.g., [17]) to define product yields for eq(33). Actual combustion chemistry is not considered in CFAST due to the complexities associated with detailed kinetics and transport.

Flame Height

CFAST includes a calculation of average flame height based on the work of Heskestad[20]. Valid for a wide range of hydrocarbon and gaseous fuels, the correlation is given by

$$H = -1.02D + 0.235 \frac{Q_f}{1000} \qquad (3.30)$$

where H is the average flame height, D the diameter of the fire and Q_f is the fire size. The mean flame height is defined as the distance from the fuel source to the top of the visible flame where the intermittency is 0.5. A flame intermittency of 0.5 means that the visible flame is above the mean 50% of the time and below the mean 50% of the time. This average flame height is included in the printed output from CFAST. Note that Q_f in Eq.(8) in Ref.[20] is in kW.

Limitation of the Algorithm for Fires and Mass Balance

CFAST depends on pyrolysis data for the source term for a fire. The usual way to obtain this data is a large-scale calorimeter, e.g., reference [[21]. Generally, a product (e.g., chair, table, bookcase) is placed under a large collection hood and ignited by a burner ($\approx$ 50 kW simulating a wastebasket) placed adjacent to the item. The combustion process then proceeds under assumed free-burning conditions, and the heat release rate is measured. Potential sources of uncertainty include measurement errors related to the instrumentation and the degree to which free-burning conditions are not achieved (e.g., radiation from the gases under the hood or from the hood itself, and restrictions in the air entrained by the object causing locally reduced oxygen concentrations affecting the combustion chemistry). There are limited experimental data for upholstered furniture which suggest that prior to the onset of flashover in a compartment, the influence of the compartment on the burning behavior of the item is small. The differences obtained from the use of different types or locations of ignition sources have not been explored. These factors are discussed in reference [22].

Where small-scale calorimeter data are used, procedures are available to extrapolate to the behavior of a full-size item. These procedures are based on empirical correlations of data which exhibit significant scatter, thus limiting their accuracy. For example, for upholstered furniture, the peak heat release rates estimated by the triangular approximation method averaged 91% (range 46% to 103%) of values measured for a group of 26 chairs with noncombustible frames, but only 63% (range 46% to 83%) of values measured for a group of 11 chairs with combustible frames [23]. Also, the triangle neglects the tails of the curve; these are the initial time from ignition to significant burning of the item, and the region of burning of the combustible frame, after the fabric and filler are consumed.

The provided data and procedures only relate directly to burning of items initiated by relatively large flaming sources. Little data are currently available for release rates under smoldering combustion, or for the high external flux and low oxygen conditions characteristic of post-flashover burning. While the model allows multiple items burning simultaneously, it does not account for the synergy of such multiple fires. Thus, for other ignitions scenarios, multiple items burning simultaneously (which exchange energy by radiation and convection), combustible interior finish, and post-flashover conditions, the model can give estimates which are often non-conservative (the actual release rates would be greater than estimated). At present, the only sure way to account for all of these complex phenomena is to conduct a full-scale compartment burn and use the pyrolysis rates directly.

Burning can be constrained by the available oxygen. However, this constrained fire is not subject to the influences of radiation to enhance its burning rate, but is influenced by the oxygen available in the compartment. If a large mass loss rate is entered, the model will follow this input until there is insufficient oxygen available for that quantity of fuel to burn in the compartment. The unburned fuel (sometimes called excess pyrolysate) is tracked as it flows out in the door jet, where it can entrain more oxygen. If this mixture is within the user-constrained flammable range, it burns in the door plume. If not, it will be tracked throughout the building until it eventually collects as unburned fuel or burns in a vent. The enthalpy released in the fire compartment and in each vent, as well as the total enthalpy released, is detailed in the output of the model. Since mass and enthalpy are conserved, the total will be correct. However, since combustion did not take place adjacent to the burning object, the actual mass burned could be lower than that specified by the user. The difference will be the unburned fuel.

An oxygen combustion chemistry scheme is employed only in constrained fires. Here user-constrained hydrocarbon ratios and species yields are used by the model to predict concentrations. A balance among hydrogen, carbon, and oxygen molecules is maintained. Under some conditions, low oxygen can change the combustion chemistry, with a resulting increase in the yields of products of incomplete combustion such as carbon monoxide. However, not enough is known about these chemical processes to build this relationship into the model at the present time. Some data exist in reports of full-scale experiments (e.g., reference [24]) which can assist in making such determinations.

3.4.2 Plumes

A plume is formed above any burning object. It acts as a pump transferring mass and enthalpy from the lower layer into the upper layer. A correlation is used to predict the amount of mass and enthalpy that is transferred. A more complete plume model would predict plume entrainment by creating a separate zone and solving the appropriate equations.

Two sources exist for moving enthalpy and mass between the layers within and between compartments. Within the compartment, the fire plume provides one source. The other source of mixing between the layers occurs at vents such as doors or windows. Here, there is mixing at the boundary of the opposing flows moving into and out of the compartment. The degree of mixing is based on an empirically-derived mixing relation. Both the outflow and inflow entrain air from the surrounding layers. The flow at vents is also modeled as a plume (called the door plume or jet), and uses the same equations as the fire plume, with two differences. First, an offset is calculated to account for entrainment within the doorway and second, the equations are modified to account for the rectangular geometry of vents compared to the round geometry of fire plumes. All plumes within the simulation entrain air from their surroundings according to an empirically-derived entrainment relation. Entrainment of relatively cool, non-smoke laden air adds oxygen to the plume and allows burning of the fuel. It also causes it to expand as the plume moves upward in the shape of an inverted cone. The entrainment in a vent is caused by bi-directional flow and results from vortices formed near a shear layer. This phenomenon is called the Kelvin-Helmholtz instability [25]. It is not exactly the same as a normal plume, so some error (not measured) arises when this entrainment is approximated by a normal plume entrainment algorithm.

While experiments show that there is very little mixing between the layers at their interface, sources of convection such as radiators or diffusers of heating and air conditioning systems, and the downward flows of gases caused by cooling at walls, will cause such mixing. These are examples of phenomena which are inconsistent with the two-zone approximation. Also, the plumes are assumed not to be affected by other flows which may occur. For example, if the burning object is near the door the strong inflow of air will cause the plume axis to lean away from the door and affect entrainment of gases into the plume. Such effects are not included in the model.

As discussed above, each compartment is divided into an upper and lower layer. At the start of the simulation, the layers in each compartment are initialized at ambient conditions and by default, the upper layer volume set to 0.001 of the compartment volume (an arbitrary, small value set to avoid the potential mathematical problems associated with dividing by zero). Other values can be set. As enthalpy and mass are pumped into the upper layer by the fire plume, the upper layer expands in volume causing the lower layer to decrease in volume and the interface to move downward. If the door to the next compartment has a soffit, there can be no flow through the vent from the upper layer until the interface reaches the bottom of that soffit. Thus in the early stages the expanding upper layer will push down on the lower layer air and force it into the next compartment through the vent by expansion.

Once the interface reaches the soffit level, a door plume forms and flow from the fire connecting doorway compartment to the next compartment is initiated. As smoke flow from the fire compartment fills the second compartment, the lower layer of air in the second compartment is pushed down. As a result, some of this air flows into the fire compartment through the lower part of the (or vent). Thus, a vent between the fire compartment and connecting compartments can have simultaneous, opposing flows of air. All flows are driven by pressure and density differences that result from temperature differences and layer depths. The key to getting the correct flow is to distribute correctly the fire and plumes mass and enthalpy between the layers.

Buoyancy generated by the combustion processes in a fire causes the formation of a plume. Such a plume can transport mass and enthalpy from the fire into the lower or upper layer of a compartment. In the present implementation, we assume that both mass and enthalpy from the fire are deposited only into the upper layer. In addition the plume entrains mass from the lower layer and transports it into the upper layer. This yields a net enthalpy transfer between the two layers.

A fire generates energy at a rate Q_f. Some fraction, χ_R, will exit the fire as radiation. The remainder, χ_C, will then be deposited in the layers as convective energy or heat additional fuel which may then pyrolyze. A buoyant plume carries this energy into the upper layer. Within CFAST, the radiative fraction, χ_R, defaults to 0.30 [26]; i.e., 30% of the fire's energy is released via radiation. For other fuels, the work of Tewarson [27], McCaffrey [28], or Koseki [29] is available for reference. The typical range for the radiative fraction is from about 0.05 to 0.4.

Plume Entrainment

All plume models used to date are based on the seminal work of Morton et al. [30]. For an extended source such as a fire, the prescription needs modification. Several studies have devised such modifications. Two of these are included in CFAST, the work of McCaffrey [31] and Heskestad [32].

McCaffrey [31] estimated the mass entrained by the fire/plume from the lower into the upper layer. This correlation divides the flame/plume into three regions as given in eq 3.48. This prescription agrees with the work of Cetegen et al. [33,34] in the intermittent regions but yields greater entrainment in the other two regions. This difference is particularly important for the initial fire since the upper layer is far removed from the fire.

$$\text{flaming:} \quad \frac{\dot{m}_e}{Q_f} = 0.011 \left[\frac{z}{Q_f^{2/5}}\right]^{0.566} \quad 0.00 \leq \left[\frac{z}{Q_f^{2/5}}\right] < 0.08$$

$$\text{intermittent:} \quad \frac{\dot{m}_e}{Q_f} = 0.026 \left[\frac{z}{Q_f^{2/5}}\right]^{0.909} \quad 0.08 \leq \left[\frac{z}{Q_f^{2/5}}\right] < 0.20 \quad (3.31)$$

$$\text{plume} \quad \frac{\dot{m}_e}{Q_f} = 0.124 \left[\frac{z}{Q_f^{2/5}}\right]^{1.895} \quad 0.20 \leq \left[\frac{z}{Q_f^{2/5}}\right]$$

McCaffrey's correlation is an extension of the common point source plume model, with a different set of coefficients for each region. These coefficients are based on experimental correlations.

Heskestad analyzed both his own data [32] and that of Zukoski [35] to develop the correlation

$$\dot{m}_e = 0.071 Q_{f,C}^{1/3} (z - z_0)^{5/3} \left[1 + 0.026 Q_{f,C}^{2/3} (z - z_0)^{-5/3}\right] \quad (3.32)$$

where $Q_{f,C}$ is the convective heat release rate of the fire and z_0 is a virtual origin for the fire plume defined as $z_0/D = -1.02 + 0.083 Q_f^{2/5}/D$, here based on the total heat release rate of the fire. Both correlations provide similar results in CFAST calculations.

Entrainment Limits

In CFAST, there is a constraint on the quantity of gas which can be entrained by a plume arising from a fire. The constraint arises from the physical fact that a plume can rise only so high for a given size of a heat source. Early in a fire, when the energy flux is very small, the plume may not have sufficient energy to reach the compartment ceiling. The correct sequence of events is for a small fire to generate a plume which does not reach the ceiling or upper layer initially. The plume entrains enough cool gas to decrease the buoyancy to the point where it no longer rises. When there is sufficient energy present in the plume, it will penetrate the upper layer. To this end the following prescription has been incorporated: for a given size fire, a limit is placed on the amount of mass which can be entrained, such that no more is entrained than would allow the plume to reach the layer interface. The result is that the interface falls at about the correct rate, although it starts a little too soon, and the upper layer temperature is overpredicted, but follows experimental data after the initial phase.

For the plume to be able to penetrate the inversion formed by a hot gas layer over a cooler gas layer, the density of the gas in the plume at the point of intersection must be less than the density of the gas in the upper layer. In practice, this places a maximum on the air entrained into the plume. From conservation of mass and enthalpy

$$\dot{m}_p = \dot{m}_f + \dot{m}_e \tag{3.33}$$

$$\dot{m}_p c_p T_p = \dot{m}_f c_p T_f + \dot{m}_e c_p T_l \tag{3.34}$$

where the subscripts p, f, e, and l refer to the plume, fire, entrained air, and lower layer, respectively.

The criterion that the density in the plume region be lower than the upper layer implies that $T_u < T_p$. Solving eq 3.34 for T_p and substituting for $\dot{m}_p$ from 3.33 yields

$$T_p = \frac{T_f \dot{m}_f + T_l \dot{m}_e}{\dot{m}_f + \dot{m}_e} > T_u \tag{3.35}$$

or

$$\dot{m}_e < \frac{T_f - T_u}{T_u - T_l} \dot{m}_f < \frac{T_f}{T_u - T_l} \dot{m}_f \tag{3.36}$$

Substituting the convective energy released by the fire, $Q_{f,C} = \dot{m}_f c_p T_f$, into eq 3.36 yields the form of the entrainment limit use in the CFAST model:

$$\dot{m}_e < \frac{Q_{f,C}}{c_p (T_u - T_l)} \tag{3.37}$$

which is incorporated into the model. It should be noted that both the plume and layers are assumed to be well mixed with negligible mixing and transport time for the plume and layers.

Plume Centerline Temperature

Used to calculate convective heat transfer to targets located directly above a fire source, CFAST includes an empirical calculation of plume centerline gas temperature taken from the work of Davis[36] using the work of Baum and McCaffrey[37] and Evans[38]. Baum and McCaffrey [37] provide an experimental correlation for plume centerline temperature consistent with plume theory. The correlation gives the excess temperature as a function of height above a fire as

$$\Delta T_p = B \left(\frac{z}{D^*}\right)^{2n-1} \tag{3.38}$$

where $D^* = \left(\frac{Q_f}{\rho_\infty c_p T_\infty \sqrt{g}}\right)^{2/5}$, Q_f is the total heat release rate, z is the height above the base of the fire, and T_∞, c_p, and ρ_∞ are the temperature, heat capacity, and density of the ambient gas at height z. Depending on the location z, this may be either the lower gas layer temperature or upper gas layer temperature. The constants B and n depend on the height above the base of the fire so that

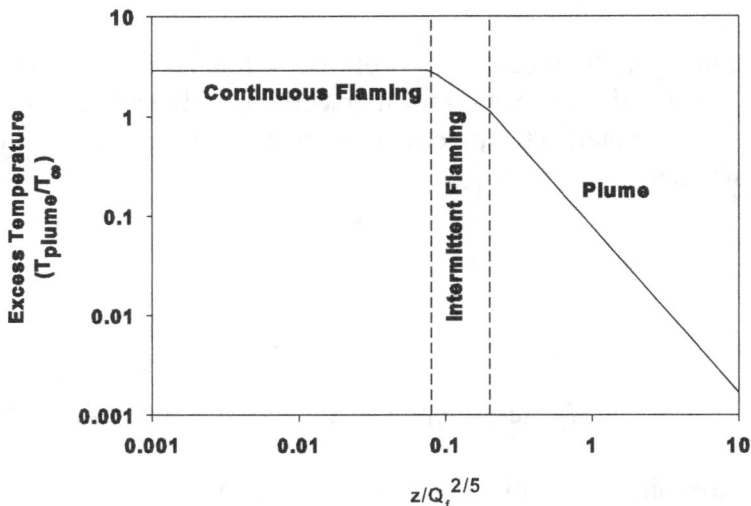

Figure 3.3: Excess plume centerline temperature from Baum and McCaffrey correlation.

flaming: $\quad \Delta T_p = 2.91 T_\infty \quad\quad 0.00 \leq z/D^* < 1.32$

intermittent: $\quad \Delta T_p = 3.81 T_\infty (z/D^*)^{-1} \quad\quad 1.32 D^* \leq z/D^* < 3.30$

plume $\quad \Delta T_p = 8.41 T_\infty (z/D^*)^{-5/3} \quad\quad 3.30 \leq z/D^*$

Figure 3.3 shows the correlation. This is used directly whenever both the fire and target location are within a single layer. When a hot layer forms so that the fire source and target location are not in the same gas layer, the correlation must be modified since the plume now includes added enthalpy by entraining hot layer gases as it moves through the upper layer to the ceiling. Evans[38] defines a virtual source and heat release rate to extend the plume into the upper layer. Evans' method defines the strength and location of the substitute source with respect to the interface between the upper and lower layers by

$$Q^*_{I,2} = \left[\frac{1 + C_T Q^{*\,2/3}_{I,1}}{\xi C_T} - \frac{1}{C_T} \right]^{3/2} \tag{3.39}$$

$$Z_{I,2} = \left[\frac{\xi Q^*_{I,1} C_T}{Q^{*\,1/3}_{I,2} \left((\xi - 1)(\beta^2 + 1) + x1 C_T Q^{*\,2/3}_{I,2} \right)} \right]^{2/5} Z_{I,1} \tag{3.40}$$

$$Q^*_{I,1} = \frac{Q_{f,C}}{\rho_\infty c_p T_\infty \sqrt{g} Z^{5/2}_{I,1}} \tag{3.41}$$

where $Z_{I,1}$ is the distance from the fire to the interface between the upper and lower gas layers, $Z_{I,2}$ is the distance from the virtual source to the layer interface, ξ is the ratio of the upper to lower

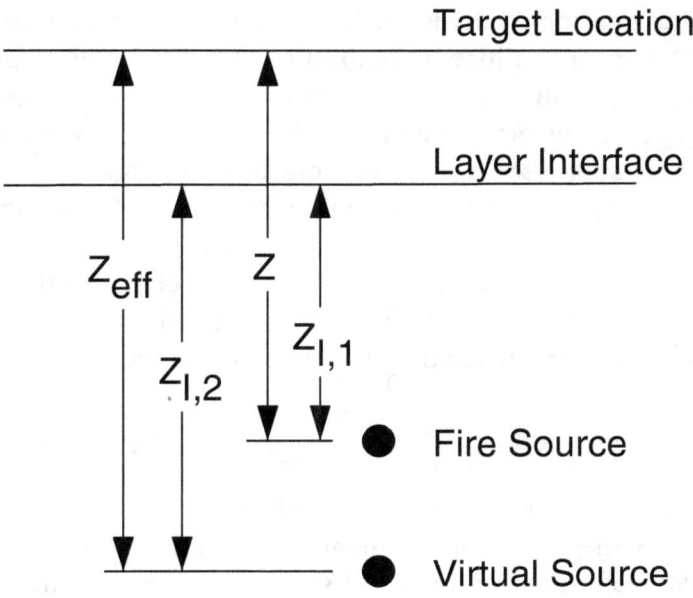

Figure 3.4: Geometry for plume centerline temperature calculation.

layer temperature, β is an experimentally determined constant [35] ($\beta^2 = 0.913$), and $C_T = 9.115$. The effective source strength and distance between the virtual source and target position is given by

$$Q_{f,C,eff} = Q^*_{I,2} \rho_\infty c_{p\infty} T_\infty \sqrt{g} Z_{I,2}^{5/2} \tag{3.42}$$

$$z_{eff} = z - Z_{I,1} + Z_{I,2} \tag{3.43}$$

(see figure 3.4). The new values of the fire source and target location are then used in the standard plume correlation where the ambient conditions are now those of the upper layer.

Limitations of the Plume Algorithms

The entrainment coefficients are empirically determined values from the work of McCaffrey[31] and Heskestad[32]. Small errors in these values will have a small effect on the fire plume or the flow in the plume of gases exiting the door of that compartment. In a multi-compartment model such as CFAST, however, small errors in each door plume are multiplicative as the flow proceeds through many compartments, possibly resulting in a significant error in the furthest compartments. The data available from validation experiments[39] discussed in the CFAST Validation Guide[8] indicate that the values for entrainment coefficients currently used in most zone models produce good agreement for a three-compartment configuration. More data are needed for larger numbers of compartments to study this further.

In real fires, smoke and gases are introduced into the lower layer of each compartment primarily due to mixing at connections between compartments and from the downward flows along walls (where contact with the wall cools the gas and reduces its buoyancy). Doorway mixing has been included in CFAST, using the same empirically derived mixing coefficients as used for calculating fire plume entrainment. Downward wall flow has not been included. This could result in under-estimates of lower layer temperatures and species concentration. Entrainment at vent (doors, windows,...) yields mixing into the lower and upper layers. The latter has been studied more extensively than the former. The door jets are not symmetric for these mixing phenomena, however. We have constrained the phenomenon for CFAST to be in the range as predicted by Zukoski et al. [40].

The plume centerline temperature is based on an experimental correlation that has been subjected to considerable validation [36]. When the desired location of the plume centerline temperature is within the flaming region of the fire, the temperature is likely to be underestimated.

3.4.3 Vent Flow

Flow through vents is a dominant component of any fire model because it is sensitive to small changes in pressure and transfers the greatest amount of enthalpy on an instantaneous basis of all the source terms (except of course for the fire and plume). Its sensitivity to environmental changes arises through its dependence on the pressure difference between compartments which can change rapidly.

CFAST models two types of vent flow, vertical flow through horizontal vents (such as ceiling holes or hatches) and horizontal flow through vertical vents (such as doors or windows). Horizontal flow is the flow which is normally thought of when discussing fires. Vertical flow is particularly important in two disparate situations: a ship, and the role of firefighters doing roof venting.

Horizontal vent flow is determined using the pressure difference across a vent. Flow at a given elevation may be computed using Bernoulli's law by first computing the pressure difference at that elevation. The pressure on each side of the vent is computed using the pressure at the floor, the height of the floor and the density.

Atmospheric pressure is about 100000 Pa. Fires produce pressure changes from 1 Pa to 1000 Pa and mechanical ventilation systems typically involve pressure differentials of about 1 Pa to 100 Pa. The pressure variables are solved to a higher accuracy than other solution variables because of the subtraction (with resulting loss of precision) needed to calculate vent flows from pressure differences.

Mass flow (in the remainder of this section, the term "flow" will be used to mean mass flow) is the dominant source term for the predictive equations because it fluctuates most rapidly and transfers the greatest amount of enthalpy on an instantaneous basis of all the source terms (except of course the fire). Also, it is most sensitive to changes in the environment. Horizontal flow encompasses flow through doors, windows and so on. Horizontal flow is discussed in section 3.4.3.1. Vertical flow occurs in ceiling vents. It is important in two separate situations: on a ship with open hatches and in house fires with roof venting. Vertical flow is discussed in section 3.4.3.2.

Flow through vents can be forced (mechanical) or natural (convective). Force flow can occur through either vertical or horizontal vents. The differences are primarily the selection rules for the source of the gases or whether the resultant plume enters the lower or upper layer of each compartment.

There is a special case of horizontal flow for long corridors. A corridor flow algorithm is incorporated to calculate the time delay from when a plume enters a compartment to when the effluent is available for flow into adjacent compartments.

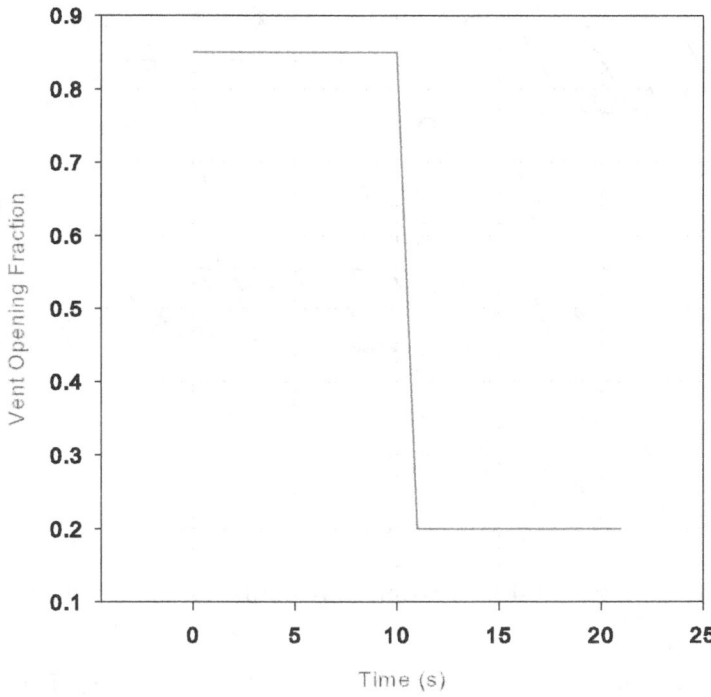

Figure 3.5: Vent opening size fraction as a function of time.

Flow through vents can be modified, that is turned on or off. This applies to the three types of vents discussed below, horizontal flow through vertical vents (HVENT), vertical flow through horizontal vents (VVENT) and forced flow (MVENT). For each keyword, there is a an initial opening fraction which is reflected in the first region in figure 3.5. This initial opening fraction can be modified by the EVENT keyword to change the fraction. This change occurs over a transition time which defaults to one second. The final fraction is the third region depicted in figure 3.5. There can be only a single transition per vent.

Horizontal Flow Through Vertical Vents

Flow through normal vents such as windows and doors is governed by the pressure difference across a vent. A momentum equation for the zone boundaries is not solved directly. Instead momentum transfer at the zone boundaries is included by using an integrated form of Euler's equation, namely Bernoulli's solution for the velocity equation. This solution is augmented for restricted openings by using flow coefficients [14] to allow for constriction from finite size doors. The flow (or orifice) coefficient is an empirical term which addresses the problem of constriction of velocity streamlines at an orifice.

Bernoulli's equation is the integral of the Euler equation and applies to general initial and final velocities and pressures. The implication of using this equation for a zone model is that the initial velocity in the doorway is the quantity sought, and the final velocity in the target compartment vanishes. That is, the flow velocity vanishes where the final pressure is measured. Thus, the

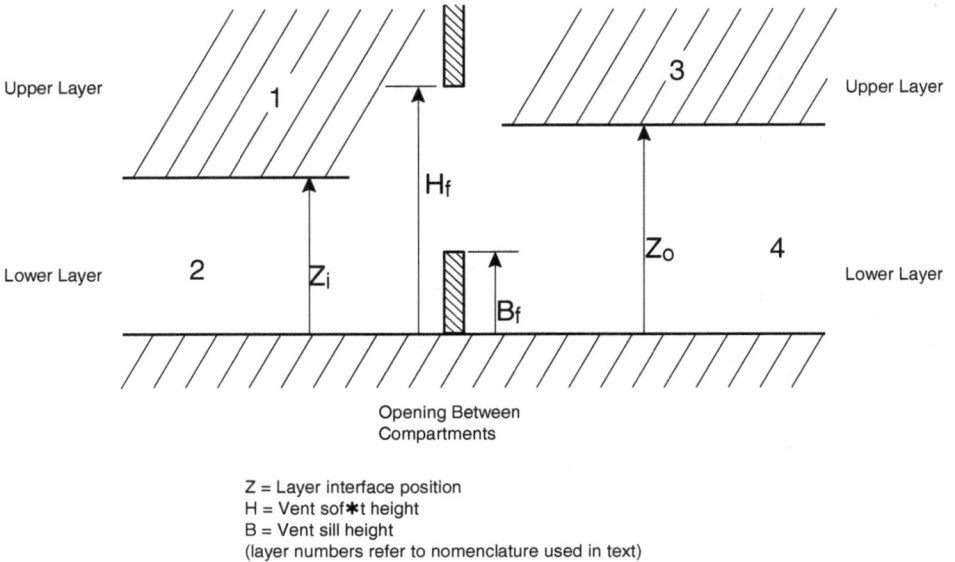

Figure 3.6: Geometry and notation for horizontal flow vents in a two-zone fire model.

pressure at a stagnation point is used. This is consistent with the concept of uniform zones which are completely mixed and have no internal flow. The general form for the velocity of the mass flow is given by

$$v = C\sqrt{\frac{2\Delta P}{\rho}} \qquad (3.44)$$

where C is the constriction (or flow) coefficient (0.7), ρ is the gas density on the source side, and ΔP is the pressure across the interface. (Note: at present we use a constant value for C for all gas temperatures).

The simplest means to define the limits of integration is with neutral planes, that is the height at which flow reversal occurs, and physical boundaries such as sills and soffits. By breaking the integral into intervals defined by flow reversal, a soffit, a sill, or a zone interface, the flow equation can be integrated piecewise analytically and then summed.

The approach to calculating the flow field is of some interest. The flow calculations are performed as follows. The vent opening is partitioned into at most six slabs where each slab is bounded by a layer height, neutral plane, or vent boundary such as a soffit or sill. The most general case is illustrated in figure 3.6.

The mass flow for each slab can be determined from

$$\dot{m}_{io} = \frac{1}{3}C(8\rho)A_{slab}\frac{x^2 + xy + y^2}{x+y} \qquad (3.45)$$

where $x = \sqrt{|P_t|}$ and $y = \sqrt{|P_b|}$. P_t and P_b are the cross-vent pressure differential at the top and bottom of the slab respectively and A_{slab} is the cross-sectional area of the slab. The value of the density, ρ, is taken from the source compartment.

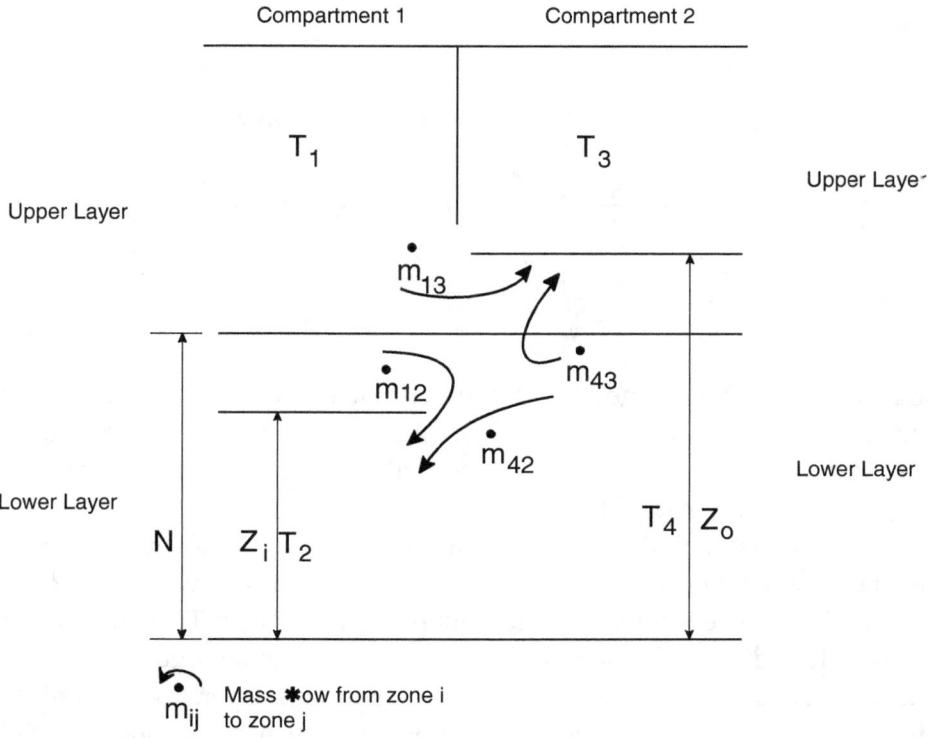

Figure 3.7: Flow patterns and layer number conventions for horizontal flow through a vertical vent.

A mixing phenomenon occurs at vents which is similar to entrainment in plumes. As hot gases from one compartment leave that compartment and flow into an adjacent compartment a door jet can exist which is analogous to a normal plume. Mixing of this type occurs for $\dot{m}_{13} > 0$ as shown in figure 3.7. To calculate the entrainment ($\dot{m}_{43}$ in this example), once again we use a plume description consistent with the work of McCaffrey, but with an extended point source. The estimate for the point source extension is given by Cetegen et al. [34]. This virtual point source is chosen so that the flow at the door opening would correspond to a plume with the heating for a equivalent doorway fire source (with respect to the lower layer) given by

$$Q_{f,eq} = c_p(T_l - T_4)\dot{m}_{13} \qquad (3.46)$$

where $Q_{f,eq}$ is the heat release rate of the doorway fire. The concept of the virtual source is that the enthalpy flux from the virtual point source should equal the actual enthalpy flux in the door jet at the point of exit from the vent using the same prescription. Thus the entrainment is calculated the same way as was done for a normal plume. The reduced height of the plume, z_p, is

$$z_p = \frac{z_{13}}{Q_{f,eq}^{2/5}} + v_p \qquad (3.47)$$

where v_p the virtual point source, is defined by inverting the entrainment process to yield

$$v_P = \frac{90.9\dot{m}^{1.76}}{Q_{f,eq}} \quad 0.00 < v_p \leq 0.08$$

$$v_P = \frac{38.5\dot{m}^{1.001}}{Q_{f,eq}} \quad 0.08 < v_p \leq 0.20 \quad (3.48)$$

$$v_P = \frac{8.10\dot{m}^{0.528}}{Q_{f,eq}} \quad 0.20 < v_p$$

Although outside of the normal range of validity of the plume model, a level of agreement with experiment is apparent (section 6 includes discussion of validation experiments for the plume model). Since a door jet forms a flat plume whereas a normal fire plume will be approximately circular, strong agreement is not expected.

The other type of mixing is much like an inverse plume and causes contamination of the lower layer. It occurs when there is flow of the type $\dot{m}_{42} > 0$. The shear flow causes vortex shedding into the lower layer and thus some of the particulates end up in the lower layer. The actual amount of mass or energy transferred is usually not large, but its effect can be large. For example, even minute amounts of carbon can change the radiative properties of the gas layer, from negligible to something finite. It changes the rate of radiation absorption significantly and invalidates the simplification of an ambient temperature lower layer. This term is predicated on the Kelvin-Helmholz flow instability and requires shear flow between two separate fluids. The mixing is enhanced for greater density differences between the two layers. However, the amount of mixing has never been well characterized. Quintiere et al. [14] discuss this phenomena for the case of crib fires in a single room, but their correlation does not yield good agreement with experimental data in the general case [41]. In the CFAST model, it is assumed that the incoming cold plume behaves like the inverse of the usual door jet between adjacent hot layers; thus we have a descending plume. It is possible that the entrainment is overestimated in this case, since buoyancy, which is the driving force, is not nearly as strong as for the usually upright plume.

Vertical Flow Through Horizontal Vents

Flow through a ceiling or floor vent can be somewhat more complicated than through door or window vents. The simplest form is uni-directional flow, driven solely by a pressure difference. This is analogous to flow in the horizontal direction driven by a piston effect of expanding gases. Once again, it can be calculated based on the Bernoulli equation, and presents little difficulty. However, in general we must deal with more complex situations that must be modeled in order to have a proper understanding of smoke movement. The first is an occurrence of puffing. When a fire exists in a compartment in which there is only one hole in the ceiling, the fire will burn until the oxygen has been depleted, pushing gas out the hole. Eventually the fire will die down. At this point ambient air will rush back in, enable combustion to increase, and the process will be repeated. Combustion is thus tightly coupled to the flow. The other case is exchange flow which occurs when the fluid configuration across the vent is unstable (such as a hotter gas layer underneath a cooler gas layer). Both of these pressure regimes require a calculation of the onset of the flow reversal mechanism.

32

Normally a non-zero cross vent pressure difference tends to drive unidirectional flow from the higher to the lower pressure side. An unstable fluid density configuration occurs when the pressure alone would dictate stable stratification, but the fluid densities are reversed. That is, the hotter gas is underneath the cooler gas. Flow induced by such an unstable fluid density configuration tends to lead to bi-directional flow, with the fluid in the lower compartment rising into the upper compartment. This situation might arise in a real fire if the room of origin suddenly had a hole punched in the ceiling. No pretense is made of being able to do this instability calculation analytically. Cooper's algorithm [42] is used for computing mass flow through ceiling and floor vents. It is based on correlations to model the unsteady component of the flow. What is surprising is that we can find a correlation at all for such a complex phenomenon. There are two components to the flow. The first is a net flow dictated by a pressure difference. The second is an exchange flow based on the relative densities of the gases. The overall flow is given by [42,43,44]

$$\dot{m} = C f(\gamma, \varepsilon) \sqrt{\frac{\Delta P}{\rho}} A_v \qquad (3.49)$$

where $\gamma = c_p/c_v$ is the ratio of specific heats, $C = 0.68 + 0.17\varepsilon$, $\varepsilon = \frac{\Delta P}{P}$, and f is a weak function of both γ and ε [42]. In the situation where we have an instability, we use Cooper's correlations for the function f. The resulting exchange flow is given by

$$\dot{m}_{ex} = 0.1 \frac{g \Delta \rho A_v^{5/2}}{\rho^{av}} \sqrt{1.0 - \frac{2A_v^2 \Delta P}{S^2 g \Delta \rho D^5}} \qquad (3.50)$$

where $D = 2\sqrt{A_v/\pi}$ and S is 0.754 for round or 0.942 for square openings, respectively [42]. Vertical flow through horizontal vents is governed by the VFLOW routines. VENTCF is the module which calculates the mass flow from one compartment to another. The values returned are $\dot{m}_{incoming}$ and $\dot{m}_{outgoing}$ through each vent. These terms are symmetric: the outgoing flow from compartment 1 to 2 is the same as incoming flow from compartment 2 to 1, though source and destination layers may be different.

The energy flux into a compartment is then determined by the relative size and temperature of the layers of the compartment from which the mass is flowing (incoming, u and l):

$$\dot{q}_{incoming} = c_p \dot{m}_u T_u + c_p \dot{m}_l T_l \qquad (3.51)$$

$$\dot{m}_u = \dot{m}_{incoming} \frac{V_u}{V} \qquad (3.52)$$

$$\dot{m}_l = \dot{m}_{incoming} \frac{V_l}{V} \qquad (3.53)$$

The mass and energy are then deposited into the upper or lower layer of the receiving compartment based on the effective temperature of the incoming flow relative to the upper and lower layers of the receiving compartment. If the temperature of the incoming flow is higher than the temperature of the lower layer, then the flow is deposited into the upper layer. This is similar to the usual plume from a fire or a doorway jet. These rules are implemented in VFLOW.

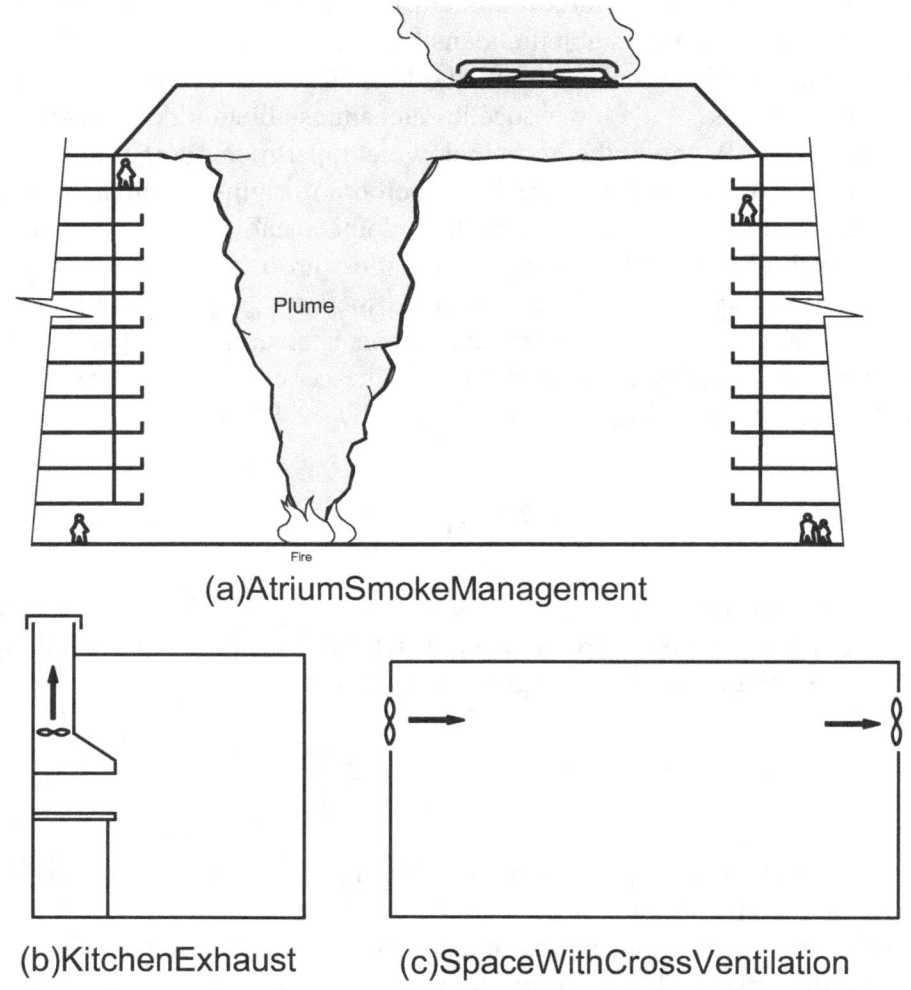

Figure 3.8: Some simple fan-duct systems.

Forced Flow

Forced flow in this version of CFAST is a supply (or exhaust) system based on constant flow through a opening/fan/opening triplet. These systems are commonly used in buildings for heating, ventilation, air conditioning, pressurization, and exhaust. Figure 3.8(a) shows smoke management by an exhaust fan at the top of an atrium, and figure 3.8(b) illustrates a kitchen exhaust. Cross ventilation, shown in figure 3.8(c), is occasionally used without heating or cooling. Generally systems that maintain comfort conditions have either one or two fans.

Further information about these systems is presented in Klote and Milke [45] and the American Society of Heating, Refrigerating and Air Conditioning Engineers (ASHRAE) [46].

This version of the model does not include ductwork or variable fans. These equations are high-order, non-linear and in some cases ill-posed, which caused a great deal of difficulty in reaching a numerical solution.

The flow through mechanical vents can be filtered. Filtering affects particulates such as smoke and the trace species. Filtering can be turned on at any time. Effectiveness is from 0% (no effect) to 100% which completely blocks flow of these two species.

3.4.4 Corridor Flow

A standard assumption in zone fire modeling is that once hot smoke enters a compartment, a well defined upper layer forms instantly throughout the compartment. This assumption breaks down in large compartments and long corridors due to the time required to fill these spaces. A simple procedure is described for accounting for the formation delay of an upper layer in a long corridor by using correlations developed from numerical experiments generated with the NIST fire model Large Eddy Simulation Model (LES), which is now the Fire Dynamic Simulation Model (FDS) [47]. FDS is a computational fluid dynamics model capable of simulating fire flow velocities and temperatures with high (≈ 0.1 m) resolution. Two parameters related to corridor flow are then estimated: the time required for a ceiling jet to travel in a corridor and the temperature distribution down the corridor. These estimates are then used in CFAST by delaying flow into compartments connected to corridors until the ceiling jet has passed these compartments.

FDS was used to estimate ceiling jet characteristics by running a number of cases for various inlet layer depths and temperatures. The vent flow algorithm in CFAST then uses this information to compute mass and enthalpy flow between the corridor and adjacent compartments. This is accomplished by presenting the vent algorithm with a one layer environment (the lower layer) before the ceiling jet reaches the vent and a two layer environment afterwards. Estimated ceiling jet temperatures and depths are used to define upper layer properties.

The problem is to estimate the ceiling jet temperature and depth as a function of time until it reaches the end of the corridor. The approach used here is to run a field model as a pre-processing step and to summarize the results as correlations describing the ceiling jet's temperatures and velocities. The steps used in this process are as follows:

1. Model corridor flow for a range of inlet ceiling jet temperatures and depths. Inlet velocities are derived from the inlet temperatures and depths.

2. For each model run calculate average ceiling jet temperature and velocity as a function of distance down the corridor.

3. Correlate the temperature and velocity distribution down the hall.

The zone fire model then uses these correlations to estimate conditions in the corridor as follows:

1. Estimate the inlet temperature, depth and velocity of the ceiling jet. If the corridor is the fire room then use a standard correlation. If the source of the ceiling jet is another room then calculate the inlet ceiling jet flow using Bernoulli's law for the vent connecting the source room and the corridor.

2. Use correlations in 3. above to estimate the ceiling jet arrival time at each vent.

3. For each vent in the corridor use lower layer properties to compute vent flow before the ceiling jet arrives at the vent and lower/upper layer properties afterwards.

Assumptions

The assumptions made in order to develop the correlations are:

- The time scale of interest is the time required for a ceiling jet to traverse the length of the corridor. For example, for a 100 m corridor with 1 m/s flow, the characteristic time would be 100 s.

- Cooling of the ceiling jet due to mixing with adjacent cool air is large compared to cooling due to heat loss to walls. Equivalently, we assume that walls are adiabatic. This assumption is conservative. An adiabatic corridor model predicts more severe conditions downstream in a corridor than a model that accounts for heat transfer to walls, since cooler ceiling jets travel slower and not as far.

- We do not account for the fact that ceiling jets that are sufficiently cooled will stagnate. Similar to the previous assumption, this assumption is conservative and results in over-predictions of conditions in compartments connected to corridors (since the model predicts that a ceiling jet may arrive at a compartment when in fact it may have stagnated before reaching it).

- Ceiling jet flow is buoyancy driven and behaves like a gravity current. The inlet velocity of the ceiling jet is related to its temperature and depth.

- Ceiling jet flow lost to compartments adjacent to the corridor is not considered when estimating ceiling jet temperatures and depths. Similarly, a ceiling jet in a corridor is assumed to have only one source.

- The temperature and velocity at the corridor inlet is constant in time.

- The corridor height and width do not effect a ceiling jet's characteristics. Two ceiling jets with the same inlet temperature, depth and velocity behave the same when flowing in corridors with different widths or heights as long as the ratio of inlet width to corridor width are equal.

- Flow entering the corridor enters at or near the ceiling. The inlet ceiling jet velocity is reduced from the vent inlet velocity by a factor of wvent/wroom where wvent and wroom are the width of the vent and room, respectively.

Corridor Jet Flow Characteristics

Ceiling jet flow in a corridor can be characterized as a one dimensional gravity current. To a first approximation, the velocity of the current depends on the difference between the density of the gas located at the leading edge of the current and the gas in the adjacent ambient air. The velocity also depends on the depth of the current below the ceiling. A simple formula for the gravity current velocity may be derived by equating the potential energy of the current, $mgd_0/2$, measured at the half-height $d_0/2$ with its kinetic energy, $mV^2/2$ to obtain

$$v = \overline{gd_0} \tag{3.54}$$

where m is mass, g is the acceleration of gravity, d_0 is the height of the gravity current and V is the velocity. When the density difference, between the current and the ambient fluid is small, the velocity, V, is proportional to $\sqrt{gd_0 \Delta\rho/\rho_{cj}} = \sqrt{gd_0 \Delta T/T_{amb}}$ where ρ_{amb} and T_{amb} are the ambient density and temperature and ρ_{cj} and T_{cj} are the density and temperature of the ceiling jet and $\Delta T = T_{cj} - T_{amb}$ is the temperature difference. Here use has been made of the ideal gas law, $\rho_{amb}T_{amb} \approx \rho_{cj}T_{cj}$. This can be shown using an integrated form of Bernoulli's law noting that the pressure drop at the bottom of the ceiling jet is $P_b = 0$, the pressure drop at the top is $P_t = gd_0(\rho_{cj} - \rho_{amb})$ and using a vent coefficient c_{vent} of 0.74, to obtain

$$
\begin{aligned}
v_0 &= C\frac{\sqrt{8}}{3}\sqrt{\frac{1}{\rho_{cj}}} \frac{P_t + \sqrt{P_t P_b} + P_b}{P_t + P_b} \\
&= C\frac{\sqrt{8}}{3} \sqrt{\frac{P_t}{\rho_{cj}}} \\
&= C\frac{\sqrt{8}}{3} \sqrt{gd_0 \frac{\rho_{amb} - \rho_{cj}}{\rho_{cj}}} \\
&\approx 0.7 \sqrt{gd_0 \frac{\Delta T}{T_{amb}}}
\end{aligned}
\quad (3.55)
$$

Formulas of the form of the above equation lead one to conclude that a ceiling jet's characteristics in a corridor depend on its depth, d_0, and relative temperature difference, $\Delta T/T_{amb}$. Therefore, as the jet cools, it slows down. If no heat transfer occurs between the ceiling jet and the surrounding walls, then the only mechanism for cooling is mixing with surrounding cool air.

Twenty numerical experiments were performed using FDS in order to better understand the effects of the inlet ceiling jet temperature and depth on ceiling jet characteristics downstream in a corridor. These cases were run with five different inlet depths and four different inlet temperatures. The inlet ceiling jet temperature rise, ΔT_0, and depth, d_0, were used to define an inlet velocity, v_0 using eq (3.55). The inlet ceiling jet depths, d_0, used in the parameter study are 0.15 m, 0.30 m, 0.45 m, 0.60 m and 0.75 m. The inlet ceiling jet temperature rises, T_0, used in the parameter study are 100 °C, 200 °C, 300 °C and 400 °C.

Correlations

Ceiling jet functions were plotted as a function of distance down a corridor for each of the 20 test cases. These results are shown in figure 3.9. Note that all but the 0.15 m ceiling jet data lie on essentially the same line.

The best fit line is given in the form of $\log \frac{\Delta T}{\Delta T_0} = a + bx$. This is equivalent to

$$
\frac{\Delta T}{\Delta T_0} = C_1 10^{bx} = C_1 \left(\frac{1}{2}\right)^{x/h_{1/2}} \quad (3.56)
$$

where $C_1 = 10^a$ and $h_{1/2} = -\log 2/b$. The parameter $h_{1/2}$ has a physical interpretation. It is the distance down the corridor where the temperature rise ΔT, falls off to 50% of its original value or equivalently, $T(x + h_{1/2}) = T(x)/2$.

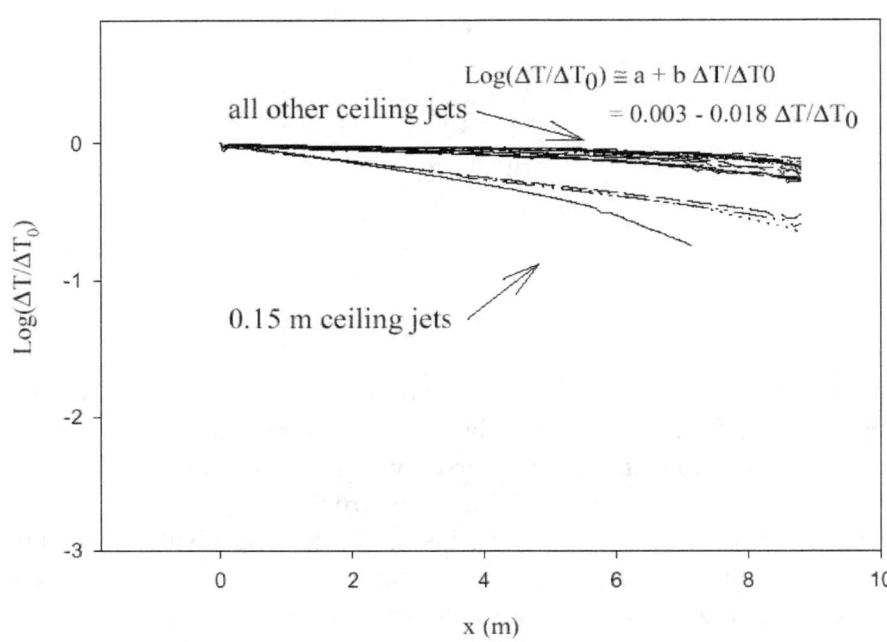

Figure 3.9: Relative excess downstream temperature in a corridor using an adiabatic temperature boundary condition for several inlet depths and inlet temperature boundary conditions. The inlet velocity is given by eq (3.55).

The half-distance, $h_{1/2}$, can be approximated by $h_{1/2} = \log 2/0.018 = 16.7$ m where $b = -0.018$ is given in figure 3.9. Similarly, the coefficient C_1 is approximated by $C_1 = 10^a = 10^{-0.003} \approx 1$ where a is determined from figure 3.9. Therefore, the temperature rise, ΔT, may be approximated by

$$\Delta T = \Delta T_0 \left(\frac{1}{2}\right)^{x/16.7} \tag{3.57}$$

The numerical experiments with FDS demonstrated that for the cases simulated, ceiling jet characteristics depend on the relative inlet temperature rise and not the inlet depth. Flow in long corridors (greater than 10 m) need to be better characterized due to the flow stagnation which may occur because of the ceiling jet's temperature decay.

3.4.5 Heat Transfer

This section discusses radiation, convection and conduction, the three mechanisms by which heat is transferred between the gas layers and the enclosing compartment walls. This section also discusses heat transfer algorithms for calculating target temperatures.

Gas layers exchange energy with their surroundings via convective and radiative heat transfer. Different material properties can be used for the ceiling, floor, and walls of each compartment (although all the walls of a compartment must be the same). Additionally, CFAST allows each surface to be composed of up to three distinct layers. This allows the user to deal naturally with the actual building construction. Material thermophysical properties are assumed to be constant, although we know that they actually vary with temperature. The users should also recognize that the mechanical properties of some materials may change with temperature, but these effects are not modeled.

Radiative transfer occurs among the fire(s), gas layers and compartment surfaces (ceiling, walls and floor). This transfer is a function of the temperature differences and the emissivity of the gas layers as well as the compartment surfaces. Typical surface emissivity values only vary over a small range. For the gas layers, however, the emissivity is a function of the concentration of species which are strong radiators, predominately smoke particulates, carbon dioxide, and water. Thus errors in the species concentrations can give rise to errors in the distribution of enthalpy among the layers, which results in errors in temperatures, resulting in errors in the flows. This illustrates just how tightly coupled the predictions made by CFAST can be.

Radiation

Radiation heat transfer forms a significant portion of the energy balance in a zone fire model, especially in the fire room. Radiative heat transfer is computed from wall and gas temperatures, emisivities and fire heat release rates. To calculate the radiation absorbed in a zone, a heat balance must be done accounting for all surfaces that radiate to and absorb radiation from a zone.

A radiation heat transfer calculation can easily dominate the computational requirements of any fire model. Approximations are then required to perform these calculations in a time consistent with other zone fire model source terms. For example, it is assumed that all zones and surfaces radiate and absorb like a gray body, that the fires radiate as point sources and that the plume does not radiate at all. Radiative heat transfer is approximated using a limited number of radiating wall surfaces, four in the fire room and two everywhere else. The use of these and other approximations allows CFAST to perform the radiation computation in a reasonably efficient manner [48].

Modeling Assumptions: The following assumptions are made in order to simplify the radiation heat exchange model used in CFAST and to make its calculation tractable.

- Iso-thermal - Each gas layer and each wall segment is assumed to be at a uniform temperature.

- Equilibrium - The wall segments and gas layers are assumed to be in a quasi-steady state. In other words, the wall and gas layer temperatures are assumed to change slowly over the duration of the time step of the associated differential equation.

- Point Soure Fires - The fire is assumed to radiate uniformly in all directions giving off a fraction, χ_R, of the total energy release rate. This radiation is assumed to originate from a single point. Radiation feedback to the fire and radiation from the plume is not modeled in the radiation exchange algorithm.

- Diffuse and gray surfaces - The radiation emitted is assumed to be diffuse and gray. In other words, the radiant fluxes emitted are independent of direction and wavelength. These assumptions allow us to infer that the emittance, ε, absorptance, α and reflectance, ρ, are related via $\varepsilon = \alpha = 1 - \rho$.

- Geometry - Rooms or compartments are assumed to be rectangular boxes. Each wall is either perpendicular or parallel to every other wall. Radiation transfer through vent openings is lost from the room.

4-Wall and 2-Wall Radiation Exchange: When computing wall temperatures, CFAST partitions a compartment into four parts; the ceiling, the floor, the wall segments above the layer interface and the wall segments below the layer interface. The radiation algorithm then computes a heat flux striking each wall segment using the surface temperature and emissivity.

The four wall algorithm used in CFAST for computing radiative heat exchange is based upon the equations developed in Siegel and Howell [49] which in turn is based on the work of Hottel [50]. Siegel and Howell model an enclosure with N wall segments and a homogeneous gas. A radiation algorithm for a two layer zone fire model requires treatment of an enclosure with two uniform gases. Hottel and Cohen [51] developed a method where the enclosure is divided into a number of wall and gas volume elements. An energy balance is written for each element. Each balance includes interactions with all other elements. Treatment of the fire and the interaction of the fire and gas layers with the walls is based upon the work of Yamada and Cooper [52]. They model fires as point heat sources radiating uniformly in all directions and use the Lambert-Beer law to model the interaction between heat emitting elements (fires, walls, gas layers) and the gas layers. By implementing a four wall rather than an N wall model, significant algorithmic speed increases are achieved. This is done by exploiting the simple structure and symmetry of the four wall problem.

The nomenclature used in this section follows that of Siegel and Howell [49]. The radiation exchange at the k'th surface is shown schematically in figure 3.10. For each wall segment k from 1 to N, a net heat flux, $\Delta \hat{q}_k$, must be found such that the energy balance,

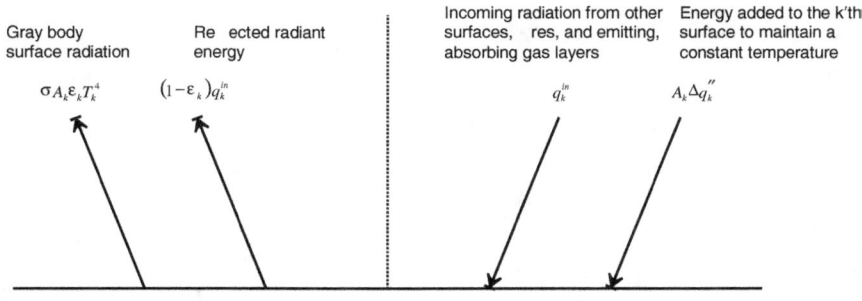

Figure 3.10: Radiation Exchange in a two-zone fire model.

$$\sigma A_k \varepsilon_k T_k^4 + (1 - \varepsilon_k) q_k^{in} = q_k^{in} + A_k \Delta q_k \quad (3.58)$$

at each wall segment k is satisfied, where σ is the Stefan-Boltzman constant, A_k is the area of the kth wall segment, ε_k is the emissivity of the kth wall segment, T_k is the temperature of the kth wall segment and q_k^{in} is the energy arriving at the kth wall segment from all other wall segments and heat sources.

Radiation exchange at each wall segment considers the emitted, reflected, incoming and net radiation terms. The unknown net radiative fluxes, Δq_k, are found by solving the modified net radiation equation

$$\Delta \hat{q}_k - \sum_{j=1}^{N} 1 - \varepsilon_j \Delta \hat{q}_j F_{k-j} \tau_{j-k} = \sigma T_k^4 - \sum_{j=1}^{N} \sigma T_k^4 F_{k-j} \tau_{k-j} - \frac{c_k}{A_k} \quad (3.59)$$

where $\Delta \hat{q}_k = \Delta q_k / \varepsilon$, F_{k-j} is the configuration factor, τ is the transmittance and other terms are previously defined.

The walls can be modeled using two surfaces or four. The four wall model is necessary for fire rooms because the temperatures of the ceiling and upper walls differ significantly. The two wall model is used for compartments that contain no fires.

To simplify the comparison between the two and four wall segment models, assume that the emissivities of all wall segments are one and that the gas absorptivities are zero. Let the room dimensions be 4 m by 4 m by 4 m, the temperature of the floor and the lower and upper walls be 300 K, and the ceiling temperature vary from 300 K to 600 K. Figure 3.11 shows a plot of the heat flux to the ceiling and upper wall as a function of the ceiling temperature [48, 53]. The two wall model predicts that the extended ceiling (a surface formed by combining the ceiling and upper wall into one wall segment) cools, while the four wall model predicts that the ceiling cools and the upper wall warms. The four-wall model moderates temperature differences that may exist between the ceiling and upper wall (or floor and lower wall) by allowing heat transfer to occur between the ceiling and upper wall. This problem does not arise when a fire is not present.

Reference [48] documents show to minimize the work required to compute the 16 configuration factors, F_{k-j}, required in a 4 wall model.

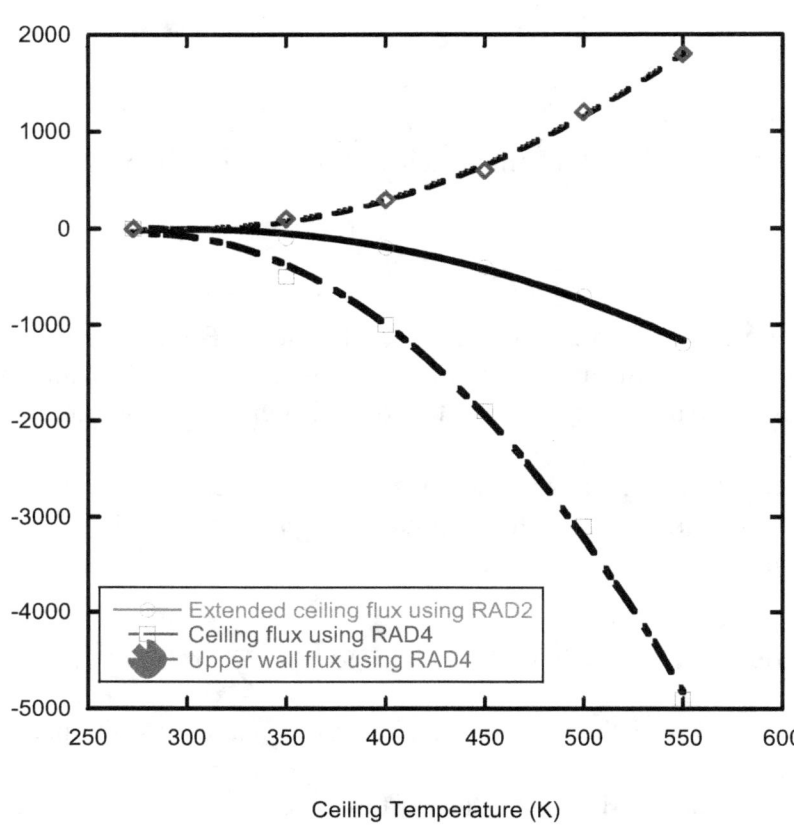

Figure 3.11: An example of the calculated two-wall (RAD2) and four-wall (RAD4) contributions to radiation exchange on a ceiling and wall surface.

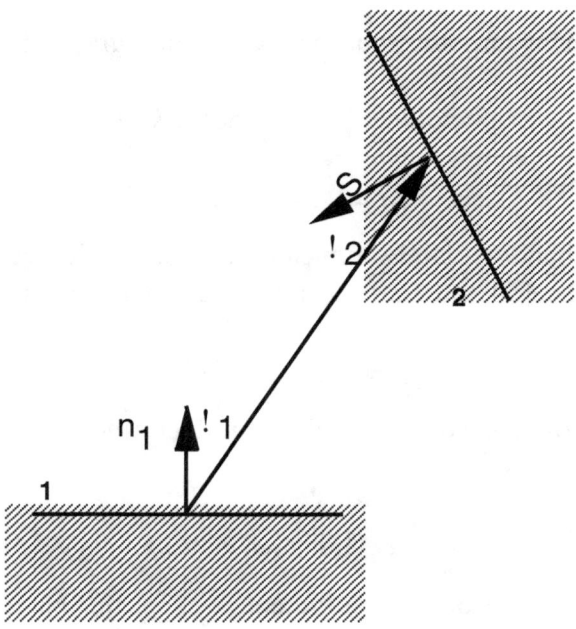

Figure 3.12: Setup for a configuration factor calculation between two arbitrarily oriented finite areas.

Configuration Factors: A configuration factor between two finite areas denoted F_{1-2} is the fraction of radiant energy given off by surface 1 that is intercepted by surface 2 and is given by

$$F_{1-2} = \frac{1}{A_1} \int_{A_1} \int_{A_2} \frac{\cos\theta_1 \theta_2}{\pi L^2} dA_2 dA_1 \qquad (3.60)$$

where L is the distance along the line of integration, θ_1 and θ_2 are the angles for surface 1 and 2 between the respective normal vectors and the line of integration, and A_1 and A_2 are the areas of the two surfaces. These terms are illustrated in figure 3.12. When the surfaces A_1 and A_2 are far apart relative to their surface area, eq (3.60) can be approximated by assuming that θ_1, θ_2 and L are constant over the region of integration to obtain

$$F_{1-2} = \frac{\cos\theta_1 \theta_2}{\pi L^2} A_2 \qquad (3.61)$$

Transmittance and Absorptance: The transmittance of a gas volume is the fraction of radiant energy that will pass through it unimpeded and is given by

$$\tau(L) = e^{-\alpha L} \qquad (3.62)$$

where α is the absorptance of the gas volume and L is a characteristic path length.

The absorptance, α, of a gas volume is the fraction of radiant energy absorbed by that volume. For a gray gas, $\alpha + \tau = 1$.

Calculating absorption for broadband gas layer radiation: In general, the transmittance and absorptance are a function of wavelength. This is an important factor to consider for the major gaseous products (CO_2 and H_2O); however soot has a continuous absorption spectrum which

43

allows the transmittance and absorptance to be approximated as "gray" [49] across the entire spectrum.

The gas absorptance, α_G, is due to the combination of the CO_2 and H_2O and is given by

$$\alpha_G = \alpha_{H_2O} + \alpha_{CO_2} - C \qquad (3.63)$$

where C is a correction for band overlap. For typical fire conditions, the overlap amounts to about half of the CO_2 absorptance [54] so the gas transmittance is approximated by

$$\tau_G = 1 - \alpha_{H_2O} - 0.5\alpha_{CO_2} \qquad (3.64)$$

The total transmittance of a gas-soot mixture is the product of the gas and soot transmittances, $\tau_T = \tau_S \tau_G$ so that

$$\tau_T = e^{-al}(1 - \alpha_{H_2O} - 0.5\alpha_{CO_2}) \qquad (3.65)$$

In the optically thin limit the absorption coefficient, a, may be replaced by the Planck mean absorption coefficient and in the optically thick limit, it may be replaced by the Rosseland mean absorption coefficient. For the entire range of optical thicknesses, Tien et al. [54] report that a reasonable approximation is $\alpha = k f_v T$ where k is a constant that depends on the optical properties of the soot particles, f_v is the soot volume fraction and T is the soot temperature in Kelvin. Values of a, have been found to be about constant for a wide range of fuels [55]. The soot volume fraction, f_v, is calculated from the soot mass, soot density and layer volume. The soot is assumed to be in thermal equilibrium with the gas layer.

Edwards' absorptance data for H_2O and CO_2 are reported [56] as log(emissivity) versus log(pressure-pathlength), with log(gas concentration) as a parameter. For each gas, these data were incorporated into a look-up table, implemented as a two-dimensional array of log(emissivity) values, with indices based on temperature and gas concentration. It is assumed that absorptance and emittance are equivalent for the gaseous species as well as for soot.

An effective path length (mean beam length, L) treats an emitting volume as if it were a hemisphere of a radius such that the flux impinging on the center of the circular base is equal to the average boundary flux produced by the real volume. The value of this radius is approximated as [55,57] $L = c4V/A$ where L is the mean beam length in meters, c is a constant (approximately 0.9, for typical geometries), V is the emitting gas volume m^3 and A is the surface area (m^2) of the gas volume. The volume and surface area are calculated from the dimensions of the layer.

For each gas, the log(absorptance) is estimated from the look-up table for that gas by interpolating both the log(temperature) and log(concentration) domains. In the event that the required absorptance lies outside the temperature or concentration range of the look-up table, the nearest acceptable value is returned. Error flags are also returned, indicating whether each parameter was in or out of range and, in the latter case, whether it was high or low. This entire process is carried out for both CO_2 and H_2O.

Computing Target Heat Flux and Temperature

The calculation of the radiative heat flux to a target is similar to the radiative heat transfer calculation discussed previously. The main difference is that CFAST does not compute feedback from

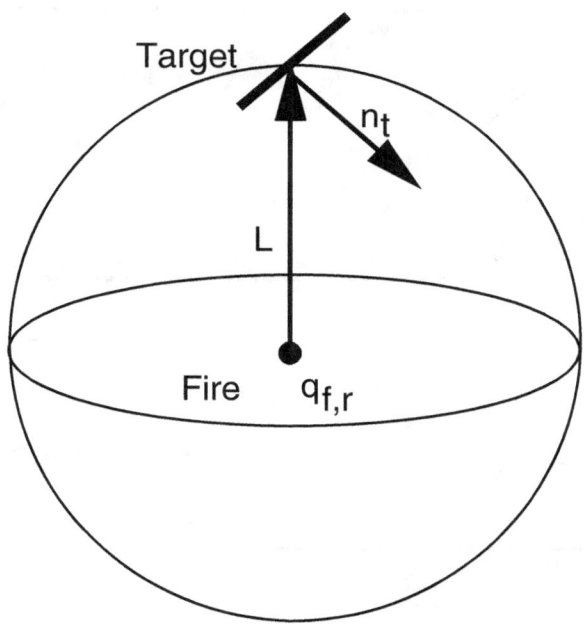

Figure 3.13: Radiative heat transfer from a point source fire to a target.

the target to the wall surfaces or gas layers. The target is simply a probe or sensor not interacting with the modeled environment.

The net flux striking a target can be used as a boundary condition in order to compute the temperature of the target. If the target is thin, then its temperature quickly rises to a level where the heat flux to and from the target are in equilibrium.

There are four components of heat flux to a target: fires, walls (including the ceiling and floor), gas layer radiation and gas layer convection.

Heat Flux from a Fire to a Target: Figure 3.13 illustrates terms used to compute heat flux from a fire to a target. Let n_t be a unit vector perpendicular to the target and θ_t be the angle between the vectors $\overline{L}$ and n_t.

Using the definition that $q_{f,r}$ is the radiative portion of the energy release rate of the fire, then the heat flux on a sphere of radius L due to this fire is $q_{f,r} / 4\pi L^2$. Correcting for the orientation of the target and accounting for heat transfer through the gas layers, the heat flux to the target is

$$q_{f,r} = \frac{q_{f,r}}{4\pi L^2} \cos(\theta_t) \tau_U(L_U) \tau_L(L_L) = -q_{f,r} \frac{n_t \overline{L}}{4\pi L^3} \tau_U(L_U) \tau_L(L_L) \quad (3.66)$$

Radiative Heat Flux from a Wall Segment to a Target: Figure 3.14 illustrates terms used to compute heat flux from a wall segment to a target. The flux, $q_{w,t}$, from a wall segment to a target can then be computed using

$$q_{w,t} = \frac{A_w q_{w,out} F_{w-t}}{A_t} \tau_U(L_U) \tau_L(L_L) \quad (3.67)$$

where $q_{w,t}$ is the flux leaving the wall segment, A_w, A_t are the areas of the wall segment and target respectively, F_{w-t} is the fraction of radiant energy given off by the wall segment that is intercepted

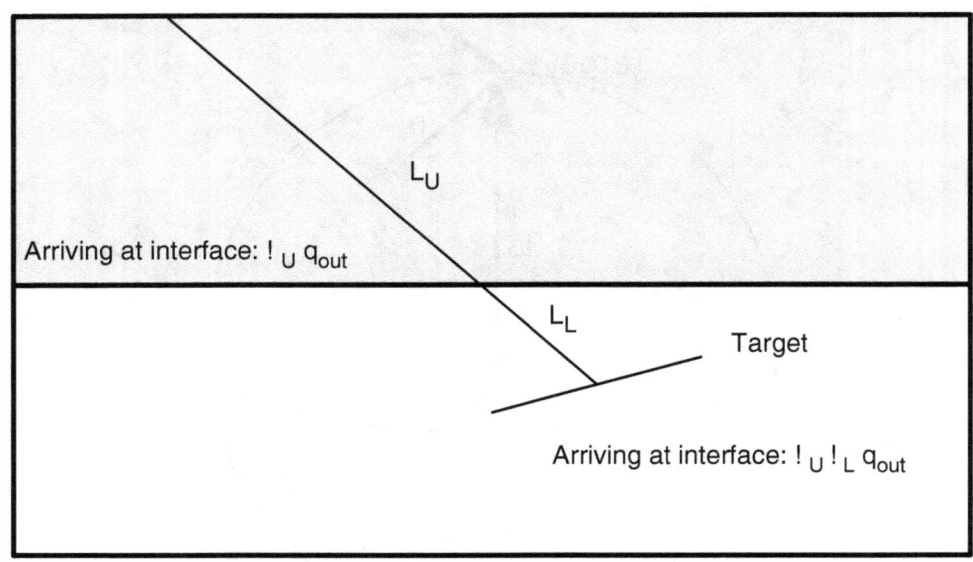

Figure 3.14: Radiative heat transfer from the upper and lower layer gas to a target in the lower layer.

by the target (i.e., a configuration factor) and $\tau_U(L_U)$ and $\tau_L(L_L)$ are defined as before. Equation (3.67) can be simplified using the symmetry relation $A_w F_{w-t} = A_t F_{t-w}$ to obtain

$$q_{w,t} = q_{w,out} F_{t-w} \tau_U(L_U) \tau_L(L_L) \tag{3.68}$$

where

$$q_{w,out} = \sigma T_w^4 - (1 - \varepsilon_w)\frac{\Delta q_w}{\varepsilon_w}, \tag{3.69}$$

T_w is the temperature of the wall segment, ε_w is the emissivity of the wall segment and Δq_w is the net flux striking the wall segment.

Radiation from the Gas Layer to the Target: Figure 3.14 illustrates the setup for calculating the heat flux from the gas layers to the target. The upper and lower gas layers in a room contribute to the heat flux striking the target if the layer absorptances are non-zero.

Let $q_{w,t,gas}$ denote the flux striking the target due to the gas in the direction of wall segment w. Then

$$q_{w,t,gas} = \begin{cases} \sigma F_{t-w}\; T_L^4 \alpha_L \tau_U + T_U^4 \alpha_U & w \text{ is in the lower layer} \\ \sigma F_{t-w}\; T_U^4 \alpha_U \tau_L + T_L^4 \alpha_L & w \text{ is in the upper layer} \end{cases} \tag{3.70}$$

The total target flux due to the gas (upper or lower layer) is obtained by summing eq (3.70) over each wall segment or

$$q_{g,t} = \sum = q_{w,t,gas} \tag{3.71}$$

Computing the Steady State Target Temperature: The steady state target temperature, T_t can be found by solving an energy balance on the target; namely

$$\varepsilon_t \sigma T_t^4 = \varepsilon_t q_{r,in} + h(T_g - T_t) \tag{3.72}$$

Note that the local gas temperature, T_g, in the convection calculation, $h(T_g - T_t)$, is taken to be either the upper layer temperature if the target is located in the upper layer, the lower layer temperature if the target is located in the lower layer, or the plume centerline temperature if the target is located directly above a fire source.

Let $f(T_t)$ be the difference between the left and right hand side of equation (3.72). Then this equation may be solved using the Newton iteration

$$T_{new} = T_{old} - \frac{f(T_{old})}{f'(T_{old})} \tag{3.73}$$

Equation (3.73) is iterated until the difference $T_{new} - T_{old}$ is sufficiently small.

Computing the Transient Target Temperature: A transient target temperature may be computed using two different methods depending on whether the target is assumed to be thin or thick. A thin target is presumed to have a constant interior temperature profile. A differential equation model may then be used to estimate the temperature rise (or fall) based upon the thermal properties of the target and the heat flux striking the front and back sides; namely

$$c \rho V \frac{dT}{dt} = A(q_f + q_b) \tag{3.74}$$

where c, ρ and V are the the specific heat, density and volume of the target, A is the cross-sectional area of the target and the two q terms are the heat flux (due to all sources) striking the front and back sides of the target.

Equation (3.74) may be solved implicitly or explicitly. When solved implicitly, the target temperature is added to the set of solution variables and equation (3.74) is added to the equation set solved by DASSL. When solved explicitly, equation (3.74) is solved as a stand-alone equation advancing the target temperature in time.

If the target is thick then it is presumed that the temperature profile within the target varies as a function of depth and therefore a partial differential equation model must be used to estimate the changing profile; namely the heat equation

$$\frac{\partial T}{\partial t} = \frac{k}{\rho C} \frac{\partial^2 T}{\partial x^2} \tag{3.75}$$

where k, ρ and C are the thermal conductivity, density and heat capasity of the target. As with the standard heat conduction model discussed later, the target heat conduction model in CFAST couples the solid to the gas phase using the relation

$$q = -k \frac{dT}{dx} \tag{3.76}$$

where q is the heat flux striking the target (again due to all sources). This equation is the statment that the flux striking the target must be consistent with the temperature gradient at the surface.

Equation (3.75) may be solved implicitly or explicitly. When solved implicitly, the target temperature is added to the set of solution variables and equation (3.76) (not equation (3.75) is added to the equation set solved by DASSL. When solved explicitly, equation (3.75) is solved as a stand-alone equation advancing the temperature profile in time.

Convection

In general, convective heat transfer, q, across a surface of area A_S, is defined as

$$q = hA_S(T_g - T_s) \tag{3.77}$$

The convective heat transfer coefficient, h, is a function of the gas properties, temperature, and velocity. The Nusselt number is defined as $Nu_L = hL/k$, which for natural convection is related to the Rayleigh number, $Ra_L = g\beta(T_s - T_g)L^3/\nu\alpha$ where L is a characteristic length of the geometry, g is the gravitational constant (m/s^2), k is the thermal conductivity (W/m^2 K), β is a volumetric expansion coefficient (K^{-1}), T_s and T_g are the temperatures of the surface and gas, respectively (K), ν is the kinematic viscosity (m^2/s), and α is the thermal diffusivity (m^2/s). All properties are evaluated at the film temperature, $T_f = (T_s + T_g)/2$. The typical correlations applicable to the problem at hand and are available in the literature [58]. The table below gives the correlations used in CFAST.

Geometry	Correlation	Restrictions
Walls	$Nu_L = \left\{ 0.825 + \dfrac{0.387 Ra_L^{1/6}}{1 + (0.492/Pr)^{9/16}}^{8/27} \right\}^2 = 0.12$	none
Ceilings and floors (hot surface up or cold surface down)	$Nu_L = 0.13 Ra_L^{1/3}$	$2x10^8 \leq Ra_L \leq 10^{11}$
Ceilings and floors (cold surface up or hot surface down)	$Nu_L = 0.16 Ra_L^{1/3}$	$2x10^8 \leq Ra_L \leq 10^{10}$

The Prandtl number, Pr, is the ratio of the kinematic viscosity and the thermal diffusivity. The thermal diffusivity, α, and thermal conductivity, k, of air are defined as a function of the film temperature from data in reference [58] as

$$\alpha = 1.0x10^{-9} T_f^{7/4} \tag{3.78}$$

$$k = \frac{0.0209 + 1.33x10^{-5} T_f}{1 - 0.000267 T_f} \tag{3.79}$$

Conduction

Procedures for solving 1-d heat conduction problems are well known, (e.g., backward difference (fully implicit), forward difference (fully explicit) or Crank-Nicolson [59]). A finite difference approach using a non-uniform spatial mesh is used to advance the wall temperature solution. The heat equation is discretized using a second order central difference for the spatial derivative and a backward differences for the time derivative. The resulting tri-diagonal system of equations is then solved to advance the temperature solution to time $t + \Delta T$. This process is repeated, using the work of Moss and Forney [60], until the heat flux striking the wall (calculated from the convection and radiation algorithms) is consistent with the flux conducted into the wall calculated via Fouriers law

$$q = -k \frac{dT}{dx} \quad (3.80)$$

where k is the thermal conductivity. This solution strategy requires a differential algebraic (DAE) solver that can simultaneously solve both differential and algebraic equations. With this method, only one or two extra equations are required per wall segment (two if both the interior and exterior wall segment surface temperatures are computed). This solution strategy is more efficient than the method of lines since fewer equations need to be solved. Wall segment temperature profiles, however, still have to be stored so there is no decrease in storage requirements. Conduction is then coupled to the room conditions by temperatures supplied at the interior boundary by the differential equation solver. The exterior boundary condition types (constant flux, insulated, or constant temperature) are specified in the configuration of CFAST.

A non-uniform mesh scheme was chosen to allow break points to cluster near the interior and exterior wall segment surfaces. This is where the temperature gradients are the steepest. A breakpoint x_b was defined by $x_b = MIN(x_p, W/2)$ where $x_p = 2\sqrt{\alpha t_{final}} \, erfc^{-1}(.05)$ and $erfc^{-1}$ denotes the inverse of the complementary error function. The value x_p is the location in a semi-infinite wall where the temperature rise is 5% after t_{final} seconds and is sometimes called the penetration depth. Eighty percent of the breakpoints were placed on the interior side of x_b and the remaining 20% were placed on the exterior side.

To illustrate the method, consider a one room case with one active wall. There are four gas equations (pressure, upper layer volume, upper layer temperature, and lower layer temperature) and one wall temperature equation. Implementation of the gradient matching method requires that storage be allocated for the temperature profile at the previous time, t, and at the next time, $t + \delta t$. Given the profile at time t and values for the five unknowns at time $t + \delta t$ (initial guess by the solver), the temperature profile is advanced from time t to time $t + \delta t$. The temperature profile gradient at $x = 0$ is computed followed by the residuals for the five equations. The DAE solver adjusts the solution variables and the time step until the residuals for all the equations are below an error tolerance. Once the solver has completed the step, the array storing the temperature profile for the previous time is updated, and the DAE solver is ready to take its next step.

Inter-compartment Heat Transfer

Heat transfer between vertically connected compartments is modeled by merging the connected surfaces for the ceiling and floor compartments or for the connected horizontal compartments. A heat conduction problem is solved for the merged walls using a temperature boundary condition for both the near and far wall. As before, temperatures are determined by the DAE solver so that the heat flux striking the walls surface (both interior and exterior) is consistent with the temperature gradient at that surface.

For horizontal heat transfer between compartments, the connections may be between partial wall surfaces, expressed as a fraction of the wall surface. CFAST first estimates conduction fractions analogous to radiative configuration factors. For example, if only one half of the rear wall in one compartment is adjacent to the front wall in a second compartment, the conduction fraction between the two compartments is 1/2. Once these fractions are determined, an average flux, q_{avg}, is calculated using

$$q_{avg} = \sum_{walls} F_{ij} q_{wall_j} \tag{3.81}$$

where F_{ij} is the fraction of flux from wall i that contributes to wall j, q_{wall_j} is the flux striking wall j.

3.4.6 Ceiling Jet

Relatively early in the development of a fire, fire-driven ceiling jets and gas-to-ceiling convective heat transfer can play a significant role in room-to-room smoke spread and in the response of near-ceiling mounted detection hardware. Cooper[61] details a model and computer algorithm to predict the instantaneous rate of convective heat transfer from fire plume gases to the overhead ceiling surface in a room of fire origin. The room is assumed to be a rectangular parallelepiped and, at times of interest, ceiling temperatures are simulated as being uniform. Also presented is an estimate of the convective heat transfer due to ceiling-jet driven wall flows. The effect on the heat transfer of the location of the fire within the room is taken into account. This algorithm has been incorporated into the CFAST model. In this section, we provide an overview of the model. Complete details are available in reference [[61].

A schematic of a fire, fire plume, and ceiling jet is shown in figure 3.15. The buoyant fire plume rises from the height Z_{fire} toward the ceiling. When the fire is below the layer interface, its mass and enthalpy flow are assumed to be deposited into the upper layer at height Z_{layer}. Having penetrated the interface, a portion of the plume typically continues to rise toward the ceiling. As it impinges on the ceiling surface, the plume gases turn and form a relatively high temperature, high velocity, turbulent ceiling jet which flows radially outward along the ceiling and transfers heat to the relatively cool ceiling surface. The convective heat transfer rate is a strong function of the radial distance from the point of impingement, reducing rapidly with increasing radius. Eventually, the relatively high temperature ceiling jet is blocked by the relatively cool wall surfaces[62]. The ceiling jet then turns downward and outward in a complicated flow along the vertical wall surfaces [63,64]. The descent of the wall flows and the heat transfer from them are eventually stopped by upward buoyant forces. They are then buoyed back upward and mix with the upper layer.

The average convective heat transfer from the ceiling jet gases to the ceiling surface, Q_{ceil}, can be expressed in integral form as

$$Q_{ceil} = \int_0^{X_{wall}} \int_0^{Y_{wall}} q_{ceil}(x,y) dx dy \tag{3.82}$$

The instantaneous convective heat flux, $q_{ceil}(x,y)$ can be determined as derived by Cooper[61] as

$$q_{ceil}(x,y) = h(T_{ad} - T_{ceil}) \tag{3.83}$$

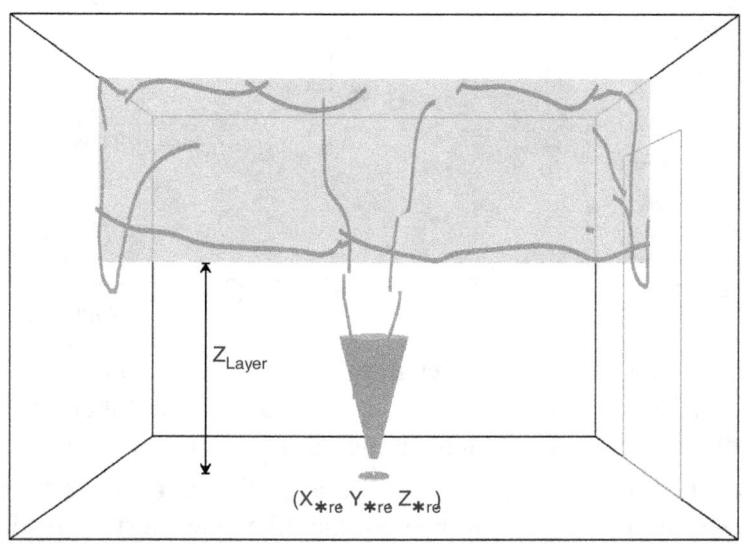

Figure 3.15: Convective heat transfer to ceiling and wall surfaces via the ceiling jet.

where T_{ad} is a characteristic ceiling jet temperature that would be measured adjacent to an adiabatic lower ceiling surface, and h is a heat transfer coefficient. h and T_{ad} are given by

$$\frac{h}{\tilde{h}} = \begin{cases} 8.82 Re_H^{-1/2} Pr^{-2/3} \left[1 - \left(5 - 0.284 Re_H^{2/5}\right) \frac{r}{H}\right] & 0 \leq \frac{r}{H} < 0.2 \\ 0.283 Re_H^{0.3} Pr^{-2/3} \left(\frac{r}{H}\right)^{-1.2} \frac{\frac{r}{H} - 0.0771}{\frac{r}{H} + 0.279} & 0.2 \leq \frac{r}{H} \end{cases} \quad (3.84)$$

$$\frac{T_{ad} - T_u}{T_u Q_H^{*2/3}} = \begin{cases} 10.22 - 14.9 \frac{r}{H} & 0 \leq \frac{r}{H} < 0.2 \\ 8.39 f\left(\frac{r}{H}\right) & 0.2 \leq \frac{r}{H} \end{cases} \quad (3.85)$$

where

$$f\left(\frac{r}{H}\right) = \frac{1 - 1.10 \left(\frac{r}{H}\right)^{0.8} + 0.808 \left(\frac{r}{H}\right)^{1.6}}{1 - 1.10 \left(\frac{r}{H}\right)^{0.8} + 2.20 \left(\frac{r}{H}\right)^{1.6} + 0.690 \left(\frac{r}{H}\right)^{2.4}} \quad (3.86)$$

$$r = \sqrt{\left(X - X_{fire}\right)^2 + \left(Y - Y_{fire}\right)^2} \quad (3.87)$$

$$\tilde{h} = \rho^u c_p g^{1/2} H^{1/2} Q_H^{*1/3}, \quad Re_H = \frac{g^{1/2} H^{3/2} Q_H^{*1/3}}{\nu^u}, \quad Q_H^* = \frac{Q}{\rho^u c_p T_u g^{1/2} H^{5/2}} \quad (3.88)$$

$$Q = \begin{cases} Q_{f,C} \dfrac{\dot{M}^*}{1+\sigma} & Z_{fire} < Z_{layer} < Z_{ceil} \\ Q_{f,C} & \begin{array}{l} Z_{fire} \geq Z_{layer} \\ Z_{layer} = Z_{ceil} \end{array} \end{cases}, \dot{M}^* = \begin{cases} 0 & -1 < \sigma \leq 0 < 0.2 \\ \dfrac{1.04599\sigma + 0.360391\sigma^2}{1+1.37748\sigma + 0.360391\sigma^2} & \sigma > 0 \end{cases} \quad (3.89)$$

$$\sigma = \dfrac{1 - \dfrac{T_u}{T_l} + C_T Q_{EQ}^{*\,2/3}}{\dfrac{T_u}{t_l}}, \quad C_T = 9.115, \quad Q_{EQ}^* = \dfrac{0.21 Q_{fc}}{c_p T_l \dot{m}_p} \quad (3.90)$$

In the above, H is the distance from the (presumed) point source fire and the ceiling, X_{fire} and Y_{fire} are the position of the fire in the room, Pr is the Prandtl number (taken to be 0.7) and ν is the kinematic viscosity of the upper layer gas which is assumed to have the properties of air and can be estimated from $\nu = 0.04128 \times 10^7 T_u^{5/2}/(T_u + 110.4)$. Q_H^* and Q_{EQ}^* are dimensionless numbers and are measures of the strength of the plume at the ceiling and the layer interface, respectively.

When the ceiling jet is blocked by the wall surfaces, the rate of heat transfer to the surface increases. Reference [61] provides details of the calculation of wall surface area and convective heat flux for the wall surfaces.

3.5 Heat Detection

Heat detection is modeled using temperatures obtained from the ceiling jet [61]. Rooms without fires do not have ceiling jets. Sensors in these types of rooms use gas layer temperatures instead of ceiling jet temperatures. The characteristic detector temperature is simply the temperature of the ceiling jet (at the location of the detector). The characteristic heat detector temperature is modeled using the differential equation [65]

$$\dfrac{dT_L}{dt} = \dfrac{\overline{v(t)}}{RTI}(T_g(t) - T_L(t)), \quad T_L(0) = T_g(0) \quad (3.91)$$

where T_L and T_g are the link and gas temperatures, v is the gas velocity, and RTI (response time index) is a measure of the sensor's sensitivity to temperature change (thermal inertia). The heat detector differential eq (3.91) may be rewritten to

$$\dfrac{dT(t)}{dt} = a(t) - b(t)T(t), \quad T(t_0) = T_0 \quad (3.92)$$

where $a(t) = \dfrac{\sqrt{V(t)}T(t)}{RTI}$ and $b(t) = \dfrac{\sqrt{V(t)}}{RTI}$. Equation (3.92) may be solved using the trapezoidal rule to obtain

$$\dfrac{T_{i+1} - T_i}{\Delta T} = \dfrac{1}{2}((a_i - b_i T_i) + (a_{i+1} - b_{i+1} T_{i+1})) \quad (3.93)$$

where the subscript i denotes time at t_i. Equation (3.93) may be simplified to

$$T_{i+1} = A_{i+1} - b_{i+1} T_{i+1} \quad (3.94)$$

where $A_{i+1} = T_i + \frac{\Delta t}{2}(a_i - b_i T_i + a_{i+1})$ and $B_{i+1} = \frac{\Delta t}{2} b_{i+1}$ which has a solution

$$T_{i+1} = \frac{A_{i+1}}{1 + B_{i+1}} = \frac{1 - \frac{\Delta t}{2} b_i}{1 + \frac{\Delta t}{2} b_{i+1}} T_i + \frac{\Delta T}{1 + \frac{\Delta t}{2} b_{i+1}} \frac{a_i + a_{i+1}}{2} \qquad (3.95)$$

Equation (3.95) reduces to the trapezoidal rule for integration when $b(t) = 0$. When $a(t)$ and $b(t)$ are constant (the gas temperature, T_g, and gas velocity, V are not changing), eq (3.91) has the solution

$$T(t) = \frac{a}{b} + \frac{e^{-b(t-t_0)}(bT_0 - a)}{b} = T_g + e^{-\frac{\sqrt{V(t)}(t-t_0)}{RTI}}(T_0 - T_g) \qquad (3.96)$$

3.6 Sprinkler Activation and Fire Attenuation

For suppression, the sprinkler is modeled using a simple model [66] generalized for varying sprinkler spray densities [67]. It is then modeled by attenuating all fires in the room where the sensor activated by a term of the form $e^{-(t-t_{act})/t_{rate}}$ where t_{act} is the time when the sensor activated and t_{rate} is a constant determining how quickly the fire attenuates. The term t_{rate} can be related to spray density of a sprinkler using a correlation developed in [67]. The suppression correlation was developed by modifying the heat release rate of a fire. For $t > t_{act}$ the heat release is given by

$$Q_f(t) = e^{-(t-t_{act})/(3Q_{spray}^{-1.8})} Q_f(t_{act}) \qquad (3.97)$$

where Q_{spray} is the spray density of a sprinkler. Note that decay rate can be formulated in terms of either the attenuation rate or the spray density. t_{rate} can be expressed in terms of Q_{spray} as $t_{rate} = 3Q_{spray}^{-1.8}$. A calculation is done to make sure that the fuel burned is consistent with the available oxygen. Once detection has occurred, then the mass and energy release rates are attenuated by

$$\dot{m}_f(t) = e^{-(t-t_{act})/t_{rate}} \dot{m}_f(t_{act}) \qquad (3.98)$$

$$Q_f(t) = e^{-(t-t_{act})/t_{rate}} Q_f(t_{act}) \qquad (3.99)$$

There are assumptions and limitations in this approach. Its main deficiency is that it assumes that sufficient water is applied to the fire to cause a decrease in the rate of heat release. This suppression model cannot handle the case when the fire overwhelms the sprinkler. The suppression model as implemented does not include the effect of a second sprinkler. Detection of all sprinklers are noted but their activation does not make the fire go out any faster. Further, multiple fires in a room imply multiple ceiling jets. It is not clear how this should be handled, i.e., how two ceiling jets should interact. When there is more than one fire, the detection algorithm uses the fire that results in the worst conditions (usually the closest fire) in order to calculate the fire sensor temperatures. Finally, the ceiling jet algorithm that we use results in temperature predictions that are warmer (for a given heat release rate) than those used in the correlation developed by Madrzykowski [68], which will cause activation sooner than expected.

3.7 Species Concentration and Deposition

CFAST uses a combustion chemistry scheme based on a carbon-hydrogen-oxygen balance. The scheme is applied in three places. The first is burning in the portion of the plume which is in the lower layer of the compartment of fire origin. The second is the portion in the upper layer, also in the compartment of origin. The third is in the vent flow which entrains air from a lower layer into an upper layer in an adjacent compartment. Included in the combustion calculation is the generation and transport of a number of species that may be produced by a fire. These species include unburned fuel, nitrogen, oxygen, carbon monoxide, carbon dioxide, hydrogen, carbon (assumed to be soot produced by the fire), hydrogen cyanide, hydrogen chloride, and an arbitrary trace species.

3.7.1 Species Transport

The species transport in CFAST is primarily a matter of bookkeeping to track individual species mass as it is generated by a fire, transported through vents, or mixed between layers in a compartment. When the layers are initialized at the start of the simulation, they are set to ambient conditions. These are the initial temperature prescribed by the user, and 23 % by mass fraction (21 % by volume fraction) oxygen, 77 % by mass fraction (79 % by volume fraction) nitrogen, a mass concentration of water prescribed by the user as a relative humidity, and a zero concentration of all other species. As fuel is burned, the various species are produced in direct relation to the mass of fuel burned (this relation is the species yields prescribed by the user for the fuel burning). Since oxygen is consumed rather than produced by the burning, the yield of oxygen is negative, and is set internally to correspond to the amount of oxygen used to burn the fuel (within the constraint of available oxygen limits discussed in sec. 3.4.1). Two special separate species calculations are included in the model, a time-integrated value for a generic toxic species, C_t, and an arbitrary trace species, TS. Both are assumed not to be part of the overall mass balance, but are rather generated by a fire and transported through a structure in a manner identical to other species.

Each unit mass of a species produced by a fire is carried in the flow to the various rooms and accumulates in the layers. The model keeps track of the mass of each species in each layer, and knows the volume of each layer as a function of time. The mass divided by the volume is the mass concentration, which along with the relative molecular mass gives the concentration in volume percent or parts per million as appropriate. Filters can be used in mechanical ventilation systems to remove species. The phenomenon has been implemented in CFAST to remove trace species and soot. It is implemented by modifying the source terms which describe gas flow. Mass that is filtered remains on the filter and is removed from the airstream. Both the resulting species density and total species removed can be analyzed. See reference [16] for an example on the use of filtering.

The calculation of radiation exchange in CFAST also depends in part on the species concentrations calculated by the model (and thus the user inputs for species yields). There are two separate radiation calculations done by CFAST. The first is for broadband radiation transfer for energy balance. The way this calculation is done is discussed in section 3.4.5. The second is a visible light calculation to answer the question of whether exit signs will be visible. The absorption of broadband radiation depends on the concentration of water, carbon dioxide and soot. The visibility calculation depends solely on the soot concentration For soot, the input for C/CO_2 is used to calculate a sooty yield from the fire (assuming all the excess carbon goes to soot). This soot generation is then transported as a species to yield a soot mass concentration to use in the optical density calculation based originally on the work of Seader and Einhorn [69]. The most recent work is by Mulholland and Croakin [70]. Based on their experimental measurements, the soot mass density is multiplied by 3,817 m^2/kg (formerly 3,500 m^2/kg) to obtain an optical density (in units of m^{-1}) which is the value reported by the model.

3.7.2 HCl Deposition

Hydrogen chloride produced in a fire can produce a strong irritant reaction that can impair escape from the fire. It has been shown [71] that significant amounts of the substance may be removed by adsorption by surfaces which contacts smoke. In our model, HCl production is treated in a manner similar to other species. However, an additional term is required to allow for deposition on, and subsequent absorption into, material surfaces.

The physical configuration that we are modeling is a gas layer adjacent to a surface (figure 3.16). The gas layer is at some temperature T_g with a concomitant density of hydrogen chloride, ρ_{HCl}. The mass transport coefficient is calculated based on the Reynolds analogy with mass and heat transfer; that is, hydrogen chloride is mass being moved convectively in the boundary layer, and some of it simply sticks to the wall surface rather than completing the journey during the convective roll-up associated with eddy diffusion in the boundary layer. The boundary layer at the wall is then in equilibrium with the wall. The latter is a statistical process and is determined by evaporation from the wall and stickiness of the wall for HCl molecules. This latter is greatly influenced by the concentration of water in the gas, in the boundary layer and on the wall itself.

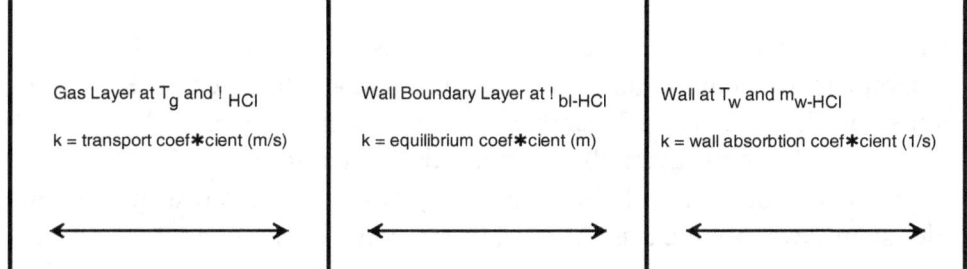

Figure 3.16: Schematic of hydrogen chloride deposition region.

The rate of addition of mass of hydrogen chloride to the gas layer is given by

$$\frac{d}{dt} m_{HCl} = source - k_c(\rho_{HCl} - \rho_{bl-HCl})A_w \qquad (3.100)$$

where source is the production rate from the burning object plus flow from other compartments. For the wall concentration, the rate of addition is

$$\frac{d}{dt} d_{w-HCl} = k_c(\rho_{HCl} - \rho_{bl-HCl}) - k_s m_{w-HCl} \qquad (3.101)$$

where the concentration in the boundary layer, ρ_{bl-HCl} is related to the wall surface concentration by the equilibrium constant k_e, by the relation $\rho_{bl-HCl} = d_{w-HCl}/k_e$. We never actually solve for

Table 3.3: Transfer coefficients for HCl deposition

Surface	b_1 (m)	b_2 (m^3/kg)	b_3 (s^{-1})	b_4 (J/gmol)	b_5 (m^3/kg)$^{b_7-b_6}$	b_6 (note a)	b_7 (note b)
Painted Gypsum	0.0063	191.8	0.0587	7476	193	1.021	0.431
PMMA	9.6×10^{-5}	0.0137	0.0205	7476	29	1.0	0.431
Ceiling Tile	4.0×10^{-3}	0.0548	0.123	7476	30^a	1.0	0.431
Cement Block	1.8×10^{-2}	5.48	0.497	7476	30^a	1.0	0.431
Calcium Silicate Board	1.9×10^{-2}	0.137	0.030	7476	30^a	1.0	0.431

a - very approximate value, insufficient data for high confidence value
b - non-dimensional

the concentration in the boundary layer, but it is available, as is a boundary layer temperature if it were of interest. The transfer coefficients are

$$k_c = \frac{\dot{q}}{\Delta T \rho_g c_p} \tag{3.102}$$

$$k_e = \frac{b_1 e^{1500/T_w}}{1 + b_2 e^{1500/T_w} \rho_{HCl}} \left[1 + \frac{b_5 (\rho_{H_2O})^{b_6}}{\rho_{H_2O,sat} - \rho_{H_2O,g}} \right] \tag{3.103}$$

$$k_s = b_3 e^{-\frac{b_4}{RT_w}} \tag{3.104}$$

The only values currently available for these quantities are shown in table 3.3 [72]. The "b" coefficients are parameters which are found by fitting experimental data to the above equations. These coefficients reproduce the adsorption and absorption of HCl reasonably well. Note though that error bars for these coefficients have not been reported in the literature.

The experimental basis for poly(methyl methacrylate) and gypsum cover a sufficiently wide range of conditions that they should be usable in a variety of practical situations. The parameters for the other surfaces do not have much experimental backing, and so their uses should be limited to comparison purposes.

3.8 Single Zone Approximation

A single zone approximation is appropriate for smoke flow far from a fire source where the two-zone layer stratification is less pronounced than in compartments near the fire. In this situation, a single zone approximation may be derived by using the normal two-zone source terms and the substitutions:

$$\begin{aligned} \dot{m}_U^{new} &= \dot{m}_L + \dot{m}_U \\ \dot{m}_L^{new} &= 0 \\ Q_U^{new} &= Q_L + Q_U \\ Q_L^{new} &= 0 \end{aligned} \tag{3.105}$$

This is used in situations where the stratification does not occur. Examples are elevators shafts, complex stairwells, natural venting ductwork, and compartments far from the fire.

3.9 Review of the Theoretical Development of the Model

Details of the software quality assurance process for CFAST is included in the Software and Model Evaluation Guide [8]. This section provides a summary of this process The current version of ASTM E1355-04 includes provisions to guide in the assessment of the theoretical basis of the model that includes a review of the model "by one or more recognized experts fully conversant with the chemistry and physics of fire phenomenon, but not involved with the production of the model. Publication of the theoretical basis of the model in a peer-reviewed journal article may be sufficient to fulfill this review [1].

CFAST has been subjected to independent review in two ways, internal and external. First, all documents issued by the National Institute of Standards and Technology receive three levels of internal review by members of the staff not involved in the preparation of the report or underlying research. The theoretical basis of CFAST is presented in this document, and is subject to internal review by staff members who are not active participants in the development of the model, but who are members of the Fire Research Division and are considered experts in the fields of fire and combustion. The same was true of previous versions of the technical reference guide over the last decade [2, 73, 74]. Externally, the theoretical basis for the model has been published in peer reviewed journals [75, 76, 77] and conference proceedings [78]. In addition, CFAST is used worldwide by fire protection engineering firms who review the technical details of the model related to their particular application. Some of these firms also publish in the open literature reports documenting internal efforts to validate the model for a particular use. Many of these studies are discussed in more detail in the present document.

In addition to the formal review, procedures were in place during the development of CFAST to assure the quality of the model. These procedures included several components:

- Review of proposed changes to the code by at least two others involved in the development process to insure that a proposed change was consistent with the rest of the CFAST code and was implemented correctly. These reviews, while informal in nature, provided a comprehensive review of the changes to the model during its development. Significant changes were documented in internal memorandums covering such areas as the numerics and structure of the model [79], improvements in the chemistry [80], convection [81], HCl deposition [82] algorithms, and output formats for the model [83]. Comparisons of the impact of the changes on the output results were often described in internal memorandums (see, for example, reference [79]).

- Internal review of the model prior to public release. In addition to the normal NIST document review process, the CFAST software was circulated internally to Fire Research Division Staff to allow interested staff members to test the model [84, 85, 86]. These memorandums detail changes to the model since the last public release of the model and provided documentation of the history of the model development.

- For each major release of CFAST, NIST has maintained a history of the source code which goes back to March 1989. While it is not practical to reconstruct the programs for each release for use with modern software tools and computer operating systems, the source code history allows the developers to examine what changes were made at each release point. This provides detailed documentation of the history of model development and is often useful to understand the impact of changes to submodels over the development of the model.

- Once a release of CFAST was approved by NIST, it was announced with a letter to model users which provided a summary of model changes and available documentation. In essence, these were a condensation of the internal memorandums, without details or printout of specific code changes. These memorandums provided documentation of the history of the model development [87,88,89,90,91].

Finally, CFAST has been reviewed and included in industry-standard handbooks such as the SFPE Handbook [92] and referenced in specific standards, including NFPA 805 [93] and NFPA 551 [94].

3.9.1 Assessment of the Completeness of Documentation

There are three primary documents on CFAST, this Technical Reference Guide, the Users Guide [7], and the Software Development and Model Evaluation Guide [8]. This document is the Technical Reference Guide and provides documentation of the governing equations, assumptions, and approximations of the various submodels. It also includes a summary description of the model structure, and numerics. The Model Users Guide includes a description of the model input data requirements and model results. The Software Development and Model Evaluation Guide describes the software quality assurance process used in the development and maintenance of the model and includes an extensive discussion of the validation of the model.

The extensive formal review process for all NIST publications in part insures the quality of the CFAST Guides. In addition, the model developers routinely receive feedback from users on the completeness of the documentation and add clarifications when needed. It is estimated that there are several thousand users of CFAST. Before new versions of the model are released, there is a "beta test" period in which the users test the new version using the updated documentation. This process is similar, although less formal, to that which most computer software programs undergo. Training courses for use of the model in fire hazard analysis have been developed from the model documentation and presented at training courses worldwide [95].

3.9.2 Assessment of Justification of Approaches and Assumptions

The technical approach and assumptions of the model have been presented in the peer reviewed scientific literature and at technical conferences. Also, all documents released by NIST are required to go through an internal editorial review and approval process. This process is designed to ensure compliance with the technical requirements, policy, and editorial quality required by NIST. The technical review includes a critical evaluation of the technical content and methodology, statistical treatment of data, uncertainty analysis, use of appropriate reference data and units, and bibliographic references. CFAST manuals are always first reviewed by a member of the Fire

Research Division, then by the immediate supervisor of the author of the document, then by the chief of the Fire Research Division, and finally by a reader from outside the division. Both the immediate supervisor and the division chief are technical experts in the field. Once the document has been reviewed, it is then brought before the Editorial Review Board (ERB), a body of representatives from all the NIST laboratories. At least one reader is designated by the Board for each document that it accepts for review. This last reader is selected based on technical competence and impartiality. The reader is usually from outside the division producing the document and is responsible for checking that the document conforms with NIST policy on units, uncertainty and scope. This reader does not need to be a technical expert in fire or combustion.

Besides formal internal and peer review, CFAST is subjected to continuous scrutiny because it is available to the general public and is used internationally by those involved in fire safety design and post fire reconstruction. The source code for CFAST is also released publicly, and has been used at various universities worldwide, both in the classroom as a teaching tool as well as for research. As a result, flaws in the theoretical development and the computer program itself have been identified and fixed. The user base continues to serve as a means to evaluate the model, which is as important to its development as the formal internal and external peer review processes.

3.9.3 Assessment of Constants and Default Values

A comprehensive assessment of the numerical parameters (such as default time step or solution convergence criteria) and physical parameters (such as empirical constants for convective heat transfer or plume entrainment) used in CFAST is not available in one document. Instead, specific parameters have been tested in various verification and validation studies performed at NIST and elsewhere. Numerical parameters are described in this Technical Reference Guide and are subject to the internal review process at NIST, but many physical parameters are extracted from the literature and do not undergo a formal review. In addition, default values for the various model inputs have been specifically reviewed by a professional fire protection engineering university professor to insure appropriate default values and suggested limits for the various input values. The model user is expected to assess the appropriateness of default values provided by CFAST and make changes to the default values if need be.

Chapter 4

Mathematical and Numerical Robustness

The mathematical and numerical robustness of a deterministic computer model depends upon three issues: the code must be transparent so that it can be understood and modified by visual inspection; it must be possible to check and verify with automated tools; and there must be a method for checking the correctness of the solution, at least for asymptotic (steady state) solutions (numerical stability and agreement with known solutions).

In order to understand the meaning of accuracy and robustness, it is necessary to understand the means by which the numerical routines are structured. In this chapter, details of the implementation of the model are presented, including the tests used to assess the numerical aspects of the model. These include

- the structure of the model, including the major routines implementing the various physical phenomena included in the model,

- the organization of data initialization and data input used by the model,

- the structure of data used to formulate the differential equations solved by the model,

- a summary of the main control routines in the model that are used to control all input and output, initialize the model and solve the appropriate differential equation set for the problem to be solved,

- the means by which the computer code is checked for consistency and correctness,

- analysis of the numerical implementation for stability and error propagation, and

- comparison of the results of the system model with simple analytical or numerical solutions.

4.1 Structure of the Numerical Routines

A methodology which is critical to verification of the model is the scheme used to incorporate physical phenomena. This is the subroutine structure discussed below. The method for incorporating new phenomena and insuring the correctness of the code was adopted as part of the consolidation of CCFM and FAST. This consolidation occurred in 1990 and has resulted in a more transparent,

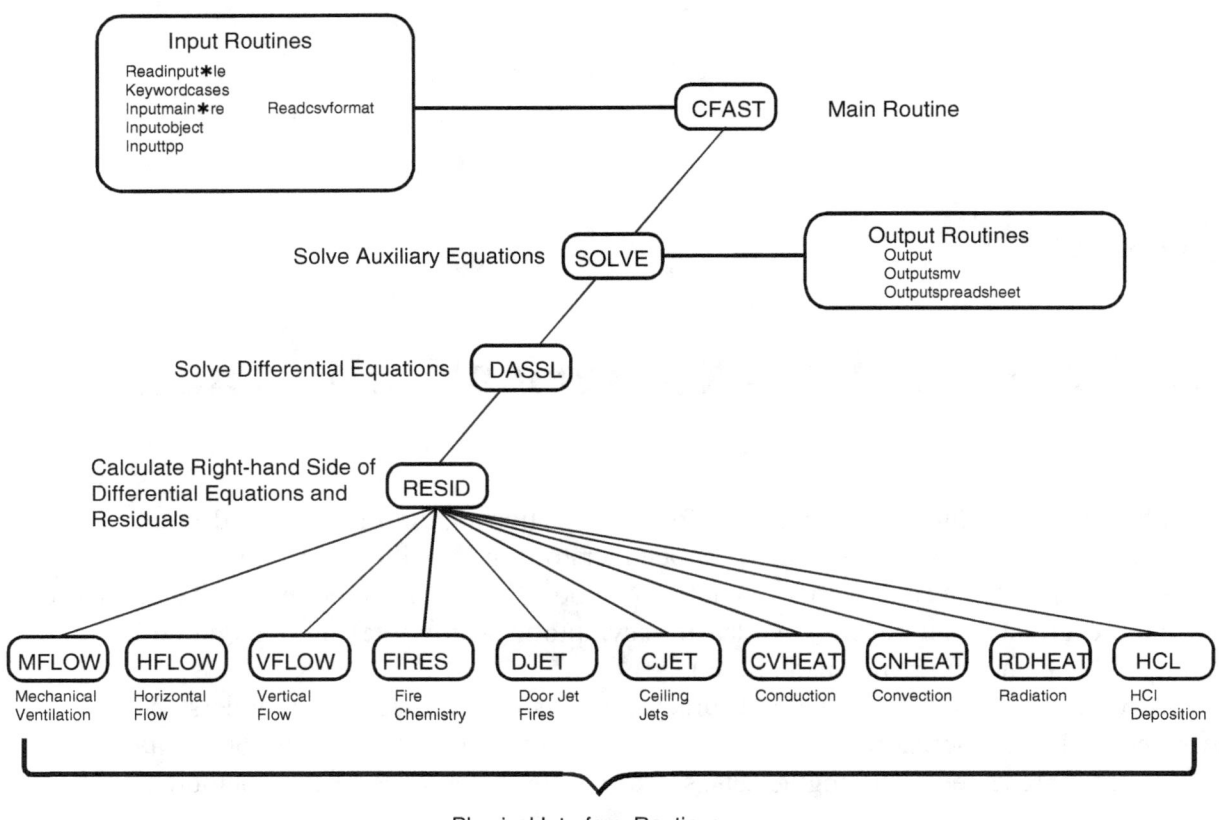

Figure 4.1: Subroutine structure for the CFAST model.

transportable and verifiable numerical model. This transparency is crucial to a verifiable and robust numerical implementation of the predictive model as discussed in the sections on code checking and numerical analysis.

The model can be split into distinct parts. There are routines for reading data, calculating results and reporting the results to a file or printer. The major routines for performing these functions are identified in figure 4.1. These physical interface routines link the CFAST model to the actual routines which calculate quantities such as mass or energy flow at one particular point in time for a given environment.

The routines SOLVE, RESID and DASSL are the key to understanding how the physical equations are solved. SOLVE is the control program that oversees the general solution of the problem. It invokes the differential equation solver DASSL [96] which in turn calls RESID to solve the transport equations. Given a solution at time t, what is the solution at time t plus a small increment of time, Δt, (where the time increment is determined dynamically by the program to insure convergence of the solution at $t + \Delta t$)? The differential equations are of the form

$$\frac{dy}{dt} = f(y, t), \quad y(t_0) = y_0 \tag{4.1}$$

where y is a vector representing pressure, layer height, mass and such, and f is a vector function that represents changes in these values with respect to time. The term y_0 is an initial condition at the initial time t_0. The subroutine RESID computes the right hand side of eq (4.1) and returns a set of residuals of that calculation to be compared to the values expected by DASSL. DASSL then checks for convergence. Once DASSL reaches an error limit (defined as convergence of the equations) for the solution at $t + \Delta t$, SOLVE then advances the solution of species concentration, wall temperature profiles, and mechanical ventilation for the same time interval. Note that there are several distinct time scales that are involved in the solution of this type of problem. The fastest will be chemical kinetics. We avoid that scale by assuming that the chemistry is infinitely fast. The next larger time scale is that associated with the flow field. These are the equations which are cast into the form of ordinary differential equations. Then there is the time scale for mechanical ventilation, and finally, heat conduction through objects. Chemical kinetic times are typically on the order of milliseconds. The transport time scale are on the order of 0.1 s. The mechanical ventilation and conduction time scales are typically several seconds, or even longer. The time step is dynamically adjusted to a value appropriate for the solution of the currently defined equation set. In addition to allowing a more correct solution to the pressure equation, very large time steps are possible if the problem being solved approaches steady-state.

4.2 Code Checking

There are two means to automate checking the correctness of the language used by a numerical model. The first is the use of standard methods for checking the structure and interface. Programs such as Flint and Lint are standard tools to do such checking. They are applied to the whole model. There are three aspects of the model checked by this procedure: correctness of the interface, undefined or incorrectly defined (or used) variables and constants, and completeness of loops and threads. It does not check for the correctness of the numerical use of constants or variables only that they are used correctly in a syntactical sense. Lint is part of most C language distributions of Unix. Flint is the equivalent for the FORTRAN language. Though it is not usually included with FORTRAN distributions Flint is generally available [1]. Both have been used with CFAST.

The second is to use a variety of computer platforms to compile and run the code. Since FORTRAN and C are implemented differently for various computers, this represents both a numerical check as well as a syntactic check. CFAST has been compiled for the Sun (Solaris), SGI (Irix), the windows-based PCs (Lahey, Digital, and Intel FORTRAN), and the Concurrent Computer platforms. Within the precision afforded by the various hardware implementations, the answers are identical [2].

4.3 Numerical Tests

There are two components to testing the numerical solutions of CFAST. First, the DASSL solver is well tested for a wide variety of differential equations, and is widely used and accepted [96]. Also, the radiation and conduction routines are tested with known solutions. These are not analytical tests, but physical limits, such as an object immersed in a fluid of constant temperature, to

---

[1] Cleanscape Software, 445 Sherman Ave, Ste. Q, Palo Alto, CA 94306
[2] Typically one part in 10^{-6}, which is the error limit used for DASSL.

which the temperature must equilibrate. The solver(s) must show that the differential equations asymptotically converge to these answers.

The second is to insure that the coupling between algorithms and the solver is correct. Most errors are avoided because of the structure discussed in section 4.1. The error due to the numerical solution is far less than that associated with the model assumptions. Two examples of this are the coupling of mechanical ventilation with buoyant flow, and the Nusselt number assumption for boundary layer convection. For the former, the coupling of a network of incompressible flow with an ODE for compressible flow has to deal with disparate calculations of pressure. For the latter, a very small time step occurs when a floor is heated and the thermal wave reaches the far (unexposed) side. This is a limitation of the physical implementation of the heat flow algorithm (convection). The solver arrives at the correct solution, but the time step becomes very small in order to achieve this.

Numerical error can be divided into three categories: roundoff, truncation and discretization error. Roundoff error occurs because computers represent real numbers using a finite number of digits. Truncation error occurs when an infinite process is replaced by a finite one. This can happen, for example, when an infinite series is truncated after a finite number of terms or when an iteration is terminated after a convergence criterion has been satisfied. Discretization error occurs when a continuous process such as a derivative is approximated by a discrete analog such as a divided difference. CFAST is designed to use 64-bit precision for real number calculations to minimize these effects.

Implicit in solving the equations discussed in chapter 3, is that the solver will arrive at a solution. Inherent in the DASSL solver are convergence criteria for the mass and energy balance within CFAST to insure mass and energy conservation within 1 part in 10^6. There are, however, limitations introduced by the algorithmic realization of physical models, that can produce errors and instabilities. Using the example above, if a mechanical ventilation system injects or removes mass and enthalpy from a small duct, then there can be a stability issue with the layer interface bobbing up and down over the duct. These are annoyances to the user community and shortcomings of the implementation of algorithm rather than failure of the system model.

Problems of this sort are noted in the frequently asked questions on the CFAST website (http://cfast.nist.gov).

4.4 Comparison with Analytic Solutions

There do not exist general analytic solutions for fire problems, even for the simplest cases. That is, there are no closed form solutions to this type of problem. However, it is possible to do two kinds of checking. The first type is discussed in the section on the theoretical basis of the model, for which individual algorithms are validated against experimental work. The second is simple experiments, especially for conduction and radiation, for which the results are asymptotic. For example, for a simple, single compartment test case with no fire, all temperatures should equilibrate asymptotically to a single value. Such comparisons are common and not usually published.

Chapter 5

Sensitivity of the Model

A sensitivity analysis considers the extent to which uncertainty in model inputs influences model output. For a sensitivity analysis, this uncertainty includes not only that inherent in the input of data for specific scenarios by the model user, but also uncertainty in empirical data or numerical parameters in the model such as the time step size used by the model to obtain a solution.

Among the purposes for conducting a sensitivity analysis are to determine

- the important variables in the models,
- the computationally valid range of values for each input variable, and
- the sensitivity of output variables to variations in input data.

Conducting a sensitivity analysis of a complex model is not a simple task and it will differ depending on the application. CFAST typically requires the user to provide numerous input parameters that describe the building geometry, compartment connections, construction materials, and description of one or more fires.

Iman and Helton[97] studied the sensitivity of complex computer models developed to simulate the risk of severe nuclear accidents which may include fire and other risks. Consistent with the work of Iman and Helton[97], ASTM E1355[1] provides overall guidance on typical areas of evaluation of the sensitivity of deterministic fire models. These areas may involve one or more of the following techniques: finite difference or direct analysis methods that provide an explicit solution of the sensitivity equations associated with the governing equations of the model, factorial design or Latin hypercube sampling studies that investigate the effect of varying the input parameters and consequential interactions between parameters that may be deemed important, and global or response surface methods that investigate the overall behavior of model outputs for a desired range of inputs.

This chapter provides a review of the sensitivity studies that have been conducted using CFAST with an emphasis on uncertainty in the input. Other sensitivity investigations of CFAST are also available [98, 99, 100].

5.1 Factorial Design Studies

Khoudja[[101] has studied the sensitivity of an early version of the FAST[2] (predecessor to CFAST) model with a fractional factorial design involving two levels of 16 different input parameters. The statistical design, taken from the texts by Box and Hunter [102], and Daniel [103] reduced the necessary model runs from more than 65000 to 256 by studying the interactions of input parameters simultaneously. The choice of values for each input parameter represented a range for each parameter. The analysis of the FAST model showed sensitivity to heat loss to the compartment walls and to the number of compartments in the simulation. Without the inclusion of surface thermophysical properties, this model treats surfaces as adiabatic for conductive heat transfer. Thus, consistent sensitivity should be expected. Sensitivity to changes in thermal properties of the surfaces were not explored.

Walker [104] discussed the uncertainties in components of zone models and showed how uncertainty within user-supplied data affects the results of calculations using CFAST as an example. The study systematically varied inputs related to the fire (heat release rate, heat of combustion, mass loss rate, radiative fraction, and species yields) and compartment geometry (vent size and ceiling height) ranging from $\pm 1\%$ to $\pm 20\%$ of base values for a one-compartment scenario. Heat release rate and ceiling height are seen to be the dominant input variables in the simulations. Upper layer temperature changed $\pm 10\%$ for a $\pm 10\%$ change in heat release rate. Typical variation of ± 10 s in time to untenable conditions for a 20% variation in the inputs was noted for the scenario studied.

Peacock et al. [98] studied the sensitivity of CFAST for a range of input parameters. They used simple factorial designs for model inputs deemed important to investigate local behavior of important model outputs along with response surface methods to evaluate overall model behavior. Results of the parametric investigations are discussed below and the application of response surface methods is summarized in section 5.2. Both are discussed in more detail in reference [98].

5.1.1 Model Inputs and Outputs

Most studies of modeling related to fire hazard and fire reconstruction present a consistent set of variables of interest to the model user [105, 106, 99, 107]: upper and lower gas layer temperatures, gas species concentrations, and layer interface position. Other variables of interest include

- mass pyrolysis and heat release rate,
- room pressure, and
- vent flow.

Although there are certainly other comparisons of interest, these will provide evidence of the sensitivity of the model to most model inputs. Tables 4 and 5 show typical inputs and outputs for the CFAST model.

Consider the following fire scenario: The building geometry (figure 5.1) includes four rooms on two floors with horizontal, vertical, and mechanical vents connecting the rooms and venting to the outdoors. The fire source in one of the rooms on the lower floor is a medium growth rate t-squared fire [108] chosen to simulate a mattress fire [23].

Sensitivity to Small Changes in Model Inputs

To investigate the sensitivity of the model, a number of simulations were conducted varying the input parameters about the base scenario discussed in the previous section. Both small ($\pm 10\%$) and

Table 5.1: Typical Inputs for a Two-Zone Fire Model

Ambient Conditions	Inside temperature and pressure. Outside temperature and pressure. Wind speed. Relative humidity
Building Geometry	Compartment width, depth, height, and surface material properties (conductivity, heat capacity, density, thickness). Horizontal Flow Vents: Height of soffit above floor, height of sill above floor, width of vent, angle of wind to vent, time history of vent openings and closings. Vertical Flow Vents: Area of vent, shape of vent. Mechanical Ventilation, Orientation of vent, Center height of vent, area of vent, length of ducts, diameter of ducts, duct roughness, duct flow coefficients, fan flow characteristics.
Fire Specification	Fire room, X, Y, Z position in room, fire area. Fire Chemistry: Molar Weight, Lower oxygen limit, heat of combustion, initial fuel temperature, gaseous ignition temperature, radiative fraction. Fire History: Mass loss rate, heat release rate, species yields for HCN, HCl, H/C, O_2/C, C/CO_2, CO/CO_2.

Table 5.2: Typical Outputs for a Two-Zone Fire Model

Environment	for each compartment: Compartment pressure and layer interface height. for each layer and compartment: Temperature, layer mass density, layer volume, heat release rate, gas concentrations (N2, O2, CO2, CO, H2O, HCl, HCN, soot optical density), radiative heat into layer, convective heat into layer, heat release rate in layer. for each vent and layer: Mass flow, entrainment, vent jet fire. for each fire: Heat release rate of fire, mass flow from plume to upper layer, plume entrainment, pyrolysis rate of fire. for each compartment surface: Surface temperatures.
Tenability	Temperature. Fractional Effective Dose (FED).

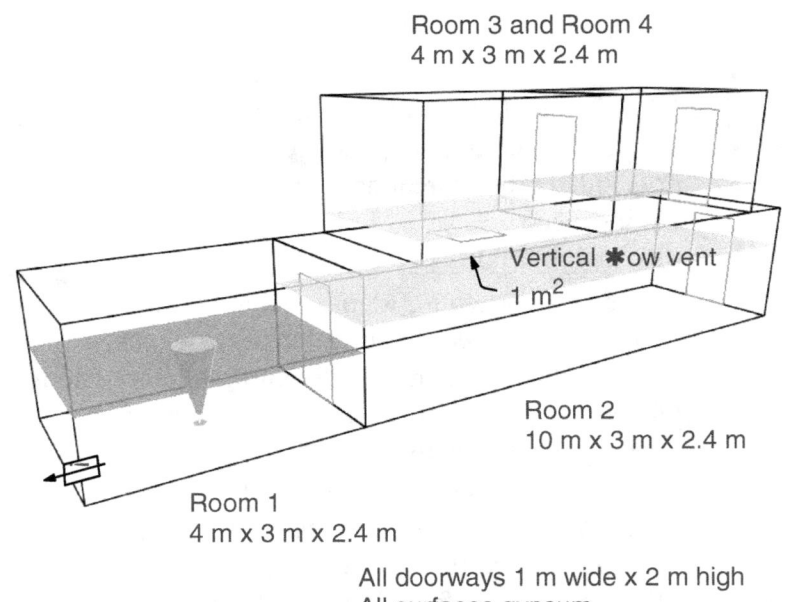

Figure 5.1: Building Geometry for base case scenario.

larger (up to an order of magnitude) variations for selected inputs were studied. Varying most of the inputs by small amounts had little effect on the model outputs. Figure 5.2 presents an example of the time dependent sensitivity of several outputs to a 10% change in room volume for the fire compartment in the scenario described above. For example, the pair of dotted-line curves labeled Upper Layer Volume were created by comparing the base case scenario with a scenario whose compartment volume was increased and decreased by 10%. The resulting curves presented on the graph are the relative difference between the variant cases and the base case defined by (Variant value - Base value)/Base value for each time point. The graph shows that temperature and pressure are insensitive to changes in the volume of the fire room since a 10% change in room volume led to smaller relative changes in layer temperature and room pressure for all times. Upper layer volume can be considered neutrally sensitive (a 10% change in room volume led to about a 10% change in layer volume). Further, this implies that there is negligible effect on the average layer interface height. This is consistent with both experimental observations in open compartment room fires and analytical solutions for single compartment steady-state fires. For transient conditions early in the fire or when the fire burns out (illustrated in the figure at 300 s when the gas burner fire heat release rate goes to zero) higher uncertainties are noted. While these are transient effects, the early phases of the fire, in particular, may be important in calculating tenability for occupants during egress. While an uncertainty in the compartment volumes results in an equivalent uncertainty in calculated outputs, accurate specification of compartment dimensions within 5% is often easily obtained.

In addition, figure 5.2 shows a somewhat constant relative difference for the changes as a function of time. As suggested by Iman and Helton [[97], an average relative difference could thus be used to characterize the model sensitivity for comparing individual inputs and outputs.

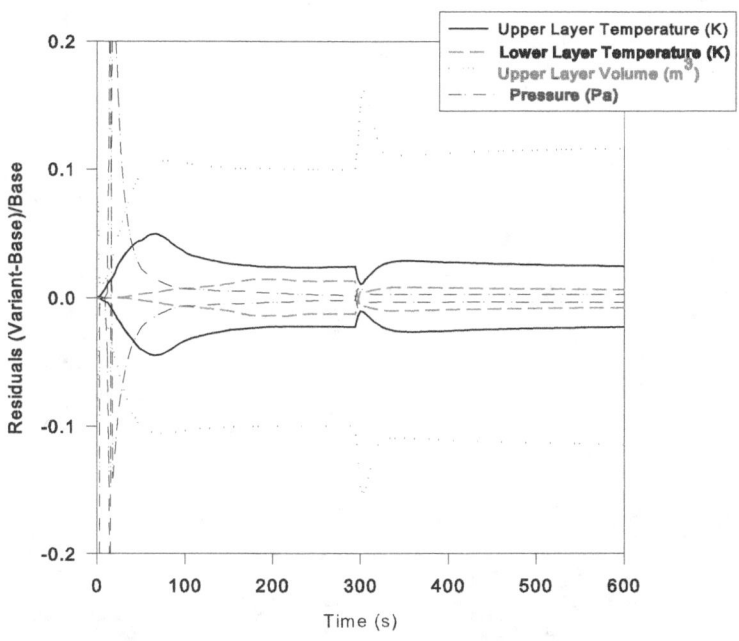

Figure 5.2: An example of time dependent sensitivity of fire model outputs to a 10% change in room volume for a single room fire scenario.

5.1.2 Sensitivity to Larger Changes in Model Inputs

To investigate the effects of much larger changes in the inputs, a series of simulations was conducted where the inputs were varied from 10% to 400% of base values. Simulations changing the heat release rate inputs from the base peak heat release rate of 750 kW are shown in figure 5.3.

Each set appears as families of curves with similar functional forms. This indicates that the heat release rate has a monotonic effect on the layer temperatures, with not as clear an effect on upper layer volume due to compartment filling and flow between compartments. Like the sensitivity to compartment volume in the previous section, changing the heat release rate by a factor of two results in a factor of two change in the upper layer temperature. Thus, in absolute terms, heat release rate and compartment volume are equally sensitive. However, compartment volume is easily determined accurately while heat release rate is typically estimated with far less accuracy and may be uncertain to within an order of magnitude or larger.

In the majority of fire cases, the most crucial question that can be asked by the person responsible for fire protection is: How big is the fire? Put in quantitative terms, this translates to: What is the heat release rate of this fire? Recently the National Institute of Standards and Technology (NIST) examined the pivotal nature of heat release rate measurements in detail [109]. Not only is heat release rate seen as the key indicator of real-scale fire performance of a material or construction, heat release rate is, in fact, the single most important variable in characterizing the flammability of products and their consequent fire hazard. Much of the remainder of this paper focuses on heat release rate as an example for examining sensitivity analysis.

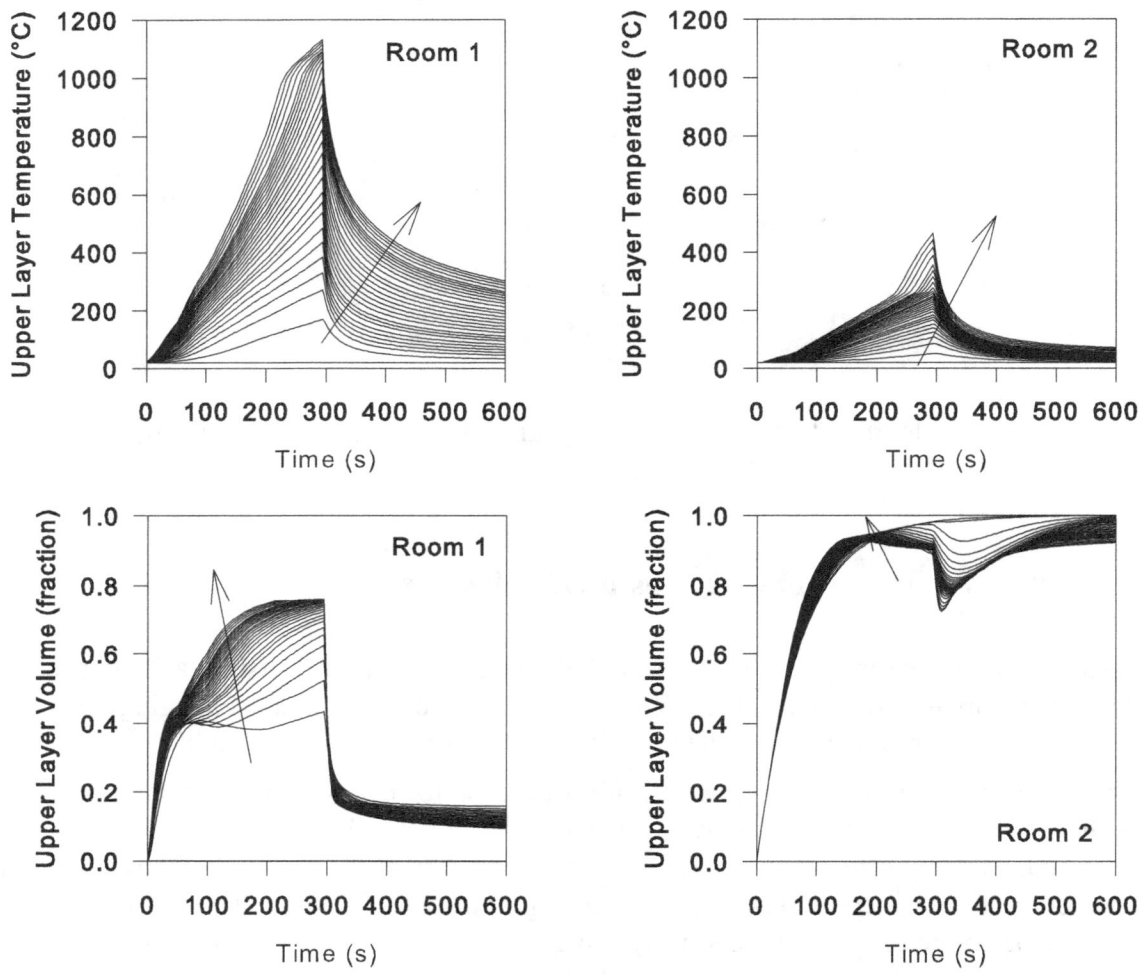

Figure 5.3: Layer temperatures and volumes in several rooms resulting from variation in heat release rate for a four-room growing fire scenario.

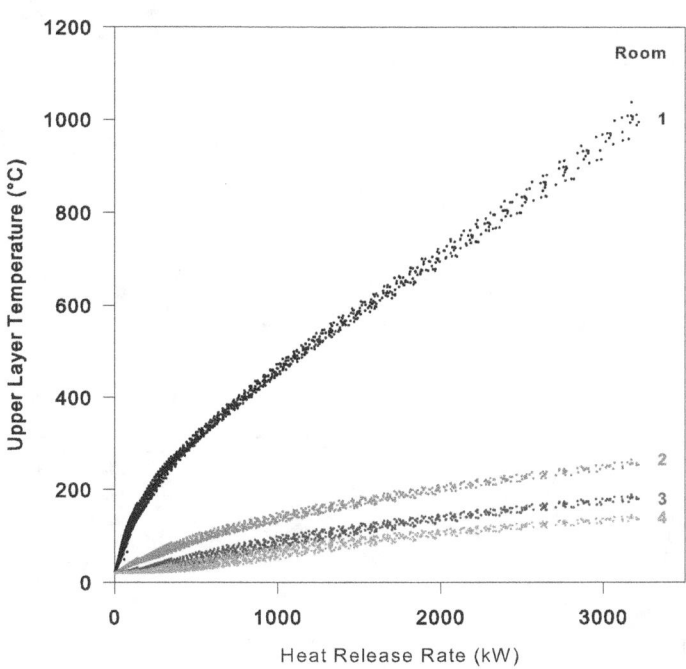

Figure 5.4: Comparison of the time dependent heat release rate and layer temperatures in several rooms for a four-room growing fire scenario.

5.2 Response Surface Studies

An extstep beyond the simple plots presented in figure 5.3 is a cross-plot of outputs of interest against the heat release rate. Figure 5.4 presents plots of the upper layer temperature (presented in figure 5.3) plotted against the heat release rate for all the simulations. The temperature curves for upper layer temperature in all four rooms (figure 5.4) show a strong functional dependence on heat release rate. Even for the wide variation in inputs, the heat release rate provides a simple predictor of the temperature in the rooms. In addition, this relationship allows calculation of the sensitivity of the temperature outputs to the heat release rate inputs as a simple slope of the resulting correlation between heat release rate and temperature.

Figure 5.5, simply a plot of the slope of the regression curves in figure 5.4, shows this sensitivity, $\partial(T)/\partial(heat\ release\ rate)$, for the four-room scenarios studied and represents all time points in all the simulations in which the peak heat release rate was varied from 0.1 to 4.0 times the base value. Except for relatively low heat release rate, the upper layer temperature sensitivity is less than 1 K/kW and usually below 0.2 K/kW. Not surprisingly, the layer that the fire feeds directly is most sensitive to changes. The lower layer in the fire room and all layers in other rooms have sensitivities less than 0.2 K/kW. This implies, for example, that if the heat release rate for a 1 MW fire is known to within 100 kW, the resulting uncertainty in the calculation of upper layer temperature in the fire room is about ± 30 K.

For upper layer volumes (figure 5.6) of both rooms 1 and 2, it is again a simple correlation between heat release rate and volume fraction (upper layer volume expressed as a fraction of the

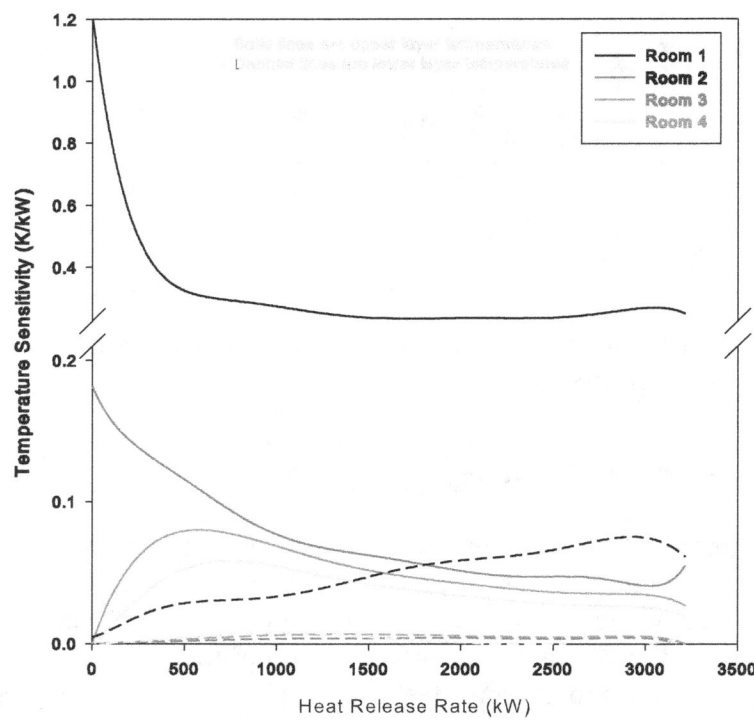

Figure 5.5: Sensitivity of temperature to heat release rate for a four-room growing fire scenario.

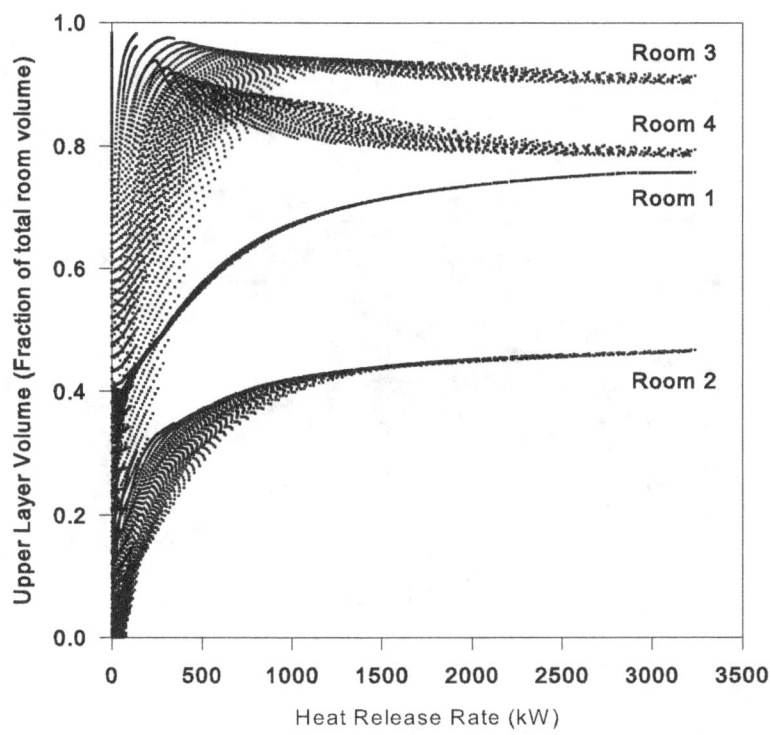

Figure 5.6: Sensitivity of temperature to heat release rate for a four-room growing fire scenario.

total room volume). The shaded gray area on the graph shows the locus of all individual time point values of temperature and volume in the four compartments of the simulation. The correlations for the upper layer volumes of room 1 and room 2 could also be differentiated as was done for the temperature correlations to obtain sensitivities for the upper layer volume. For rooms 3 and 4, the relationship is not as clear. The flow into the layers of these rooms is more complicated than for rooms 1 and 2, resulting from flow from the first floor through a vent in the floor of room 3 and from a vent to the outside in room 4. However, even these rooms approach a constant value for higher heat release rate values, implying near zero sensitivity for high heat release rate.

Figure 5.7 presents the effect of both peak heat release rate and vent opening (in the fire room) on the peak upper layer temperature. In this figure, actual model calculations, normalized to the base scenario values are indicated by circles overlaid on a surface grid generated by a spline interpolation between the data points. At high heat release rate and small vent openings, the fire becomes oxygen limited and the temperature trails off accordingly, but for the most part, the behavior of the model is monotonic in nature. Although more laborious, the approaches used to calculate sensitivities for single variable dependencies illustrated earlier are thus equally applicable to multivariate analyses.

From the surface, it is clear that heat release rate has more of an effect on the peak temperature than does the vent width. Until the fire becomes oxygen limited, the trends evident in the surface are consistent with expectations temperature goes up with rising heat release rate and down with rising vent width. The effects are not, of course, linear with either heat release rate or vent opening.

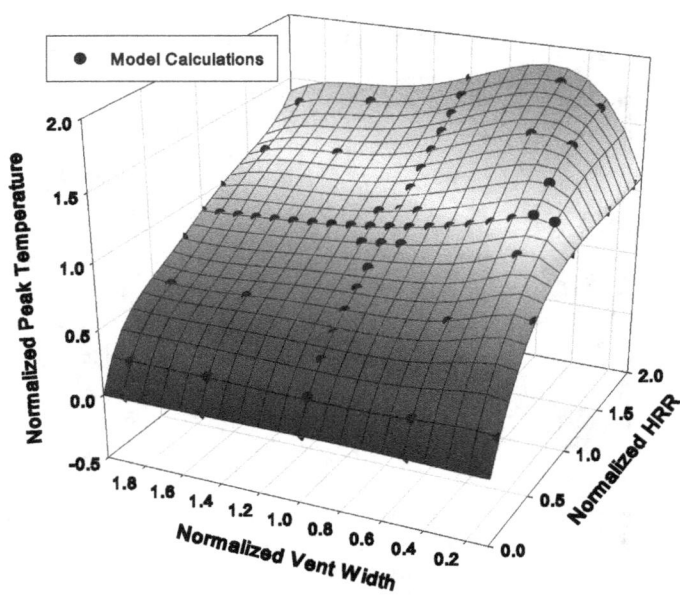

Figure 5.7: Sensitivity of temperature to heat release rate for a four-room growing fire scenario.

Plume theory and typically used algorithms for estimating upper layer temperature in a single room with a fire suggest that the dependence is on the order of $Q_f^{2/3}$ for heat release rate and $A\sqrt{h}$ for the vent opening where A is the area of the vent and h is the height of the vent. Although these correlations are based on a simple analysis of a single room fire, the dependence suggested is similar to that illustrated in figure 5.7.

5.3 Latin Hypercube Sampling Studies

Notarianni[100] developed an iterative methodology for the treatment of uncertainty in fire-safety engineering calculations to identify important model parameters for detailed study of uncertainty. She defines a nine-step process to identify crucial model inputs and parameters, select sampling methods appropriate for the important parameters, and evaluate the sensitivity of the model to chosen outcomes. Both factorial designs and Latin hypercube sampling are included in a case study involving the CFAST model. In a performance-based design of a 16 story residential structure, the impact of model uncertainty on a chosen design and inclusion of residential sprinklers in the design would effect the resulting safety of the design. For a seven-compartment scenario representing one living unit in the structure, distributions of input variables based on Latin hypercube sampling of selected ranges of the inputs were developed and used as input for a series of 500 CFAST simulations for the scenario. The results of the calculations are presented in a series of cumulative distribution functions which show the probability that a chosen criterion of the design is exceeded within a given time. Depending on the evaluation criterion chosen, times to unacceptable designs varied by as little as 10 s to as much as 470 s. To determine important input variables, Notarianni used a multivariate correlation of the input and output variables to determine statistical significance at a 95 % confidence level. Input variables deemed important in the analysis included fire-related inputs (growth rate, heat of combustion, position of the base of the fire, and generation rates of products of combustion) and door opening sizes. Other inputs were determined to be less important.

5.4 Summary

Many of the outputs of the CFAST model are quite insensitive to uncertainty in the input parameters for a broad range of scenarios. Not surprisingly, heat release rate was consistently seen as the most important variable in a range of simulations. Heat release rate and related variables such as heat of combustion or generation rates of products of combustion provide the driving force for fire-driven flows. For CFAST, all of these are user inputs. Thus, careful selection of these fire related variables are necessary for accurate predictions. Other variables related to compartment geometry such as compartment height or vent sizes, while deemed important for the model outputs, are typically more easily defined for specific design scenarios than fire related inputs. For some scenarios, such as typical building performance design, these events may need to include the effects of leakage to insure accurate predictions. For other scenarios, such as shipboard use or nuclear power facilities, leakage (or lack thereof) may be easily defined and may not be an issue in the calculations.

Chapter 6

Survey of Past Validation Work

CFAST has been subjected to extensive validation studies by NIST and others. There are two ways of comparing predictive capability with actual events. The first is simply graphing the time series curves of model results with measured values of variables such as temperature. Another approach is to consider the time to critical conditions such as flashover. Making such direct comparisons between theory and experiment provides a sense of whether predictions are reasonable. This chapter provides a review of CFAST validation efforts by NIST and others to better understand the quality of the predictions by the model.

Some of the work has been performed at NIST, some by its grantees and some by engineering firms using the model. Because each organization has its own reasons for validating the model, the referenced papers and reports do not follow any particular guidelines. Some of the works only provide a qualitative assessment of the model, concluding that the model agreement with a particular experiment is "good" or "reasonable." Sometimes, the conclusion is that the model works well in certain cases, not as well in others. These studies are included in the survey because the references are useful to other model users who may have a similar application and are interested in qualitative assessment. It is important to note that some of the papers point out flaws in early releases of CFAST that have been corrected or improved in more recent releases. Some of the issues raised, however, are still subjects of active research. Continued updates for CFAST are greatly influenced by the feedback provided by users, often through publication of validation efforts.

6.1 Comparisons with Full-Scale Tests Conducted Specifically for the Chosen Evaluation

Several studies have been conducted specifically to validate the use of CFAST in building performance design. Dembsey[110] used CFAST version 3.1 to predict the ceiling jet temperatures, surface heat fluxes and heat transfer coefficients for twenty compartment fire experiments in a compartment that is similar in size, geometry, and construction to the standard fire test compartment specified in the Uniform Building Code[111] [1]. Results from 330 kW, 630 kW, and 980 kW fires were used. In general, CFAST made predictions which were higher than the experimental

---
[1] The 1997 Uniform Building Code has been superceded by the International Building Code, 2003 Edition, International Code Council, Country Club Hills, Illinois.

results. In these cases, the temperature prediction is typically 20% to 30% higher than measured values. Much of this can be attributed to not knowing the species production (soot) and relative absorption of radiation by the gas layers which highlights the importance of scenario specification. This is the most common cause of overprediction of temperature by CFAST. A secondary source of discrepancy is correcting for radiation from thermocouple beads. The authors provide for this correction, but the corrections cited are not as large as has been reported in other fire experiments [112].

He et al. [113] describe a series of full-scale fire experiments that were designed to investigate the validity of two zone models including CFAST version 3.1. The experiments, involving steady state burning rates and a number of ventilation conditions, were conducted in a four-story building. Temperature, pressure, flow velocity, smoke density and species concentrations were measured in various parts of the building. The stack effect and its influence on temperature distribution in a stairshaft were observed. Comparisons were then made between the experimental results and the model predictions. Early in the fire there is a few percent difference [2] between the predictions and measurements; beyond 10 min, there are significant variations. Both the experiment and the model are internally consistent; that is, higher flow leads to a higher interface height (figure 13 in the paper). Once again, the difference is about 25%. The authors discuss the effect of fuel composition and correction for radiation from thermocouple beads but did not draw firm conclusions based on their measurements of fuel products.

A series of experimental results for flaming fires, obtained using realistic fires in a prototype apartment building were performed by Luo et al. [114]. Fuel configurations in the fire test included a horizontal plain polyurethane slab, mock-up chair (polyurethane slabs plus a cotton linen cover), and a commercial chair. CFAST version 3.1 typically over-predicted upper layer temperatures by 10% to 50% depending on the test conditions and measurement location in that test. The predicted and experimental time dependent upper layer temperatures were similar in shape. The time to obtain peak upper layer temperatures was typically predicted to within 15% of the experimental measurements. The authors concluded that CFAST was conservative in terms of life safety calculations.

In order to optimize fire service training facilities, the best use of resources is imperative. The work reported by Poole et al. [115] represents one aspect of a cooperative project between the city of Kitchener Fire Department (Canada) and the University of Waterloo aimed at developing design criteria for the construction of a firefighter training facility. One particular criterion is that realistic training with respect to temperature, heat release and stratification be provided in such a facility. The purpose of this paper was to compare existing analytical heat release and upper and lower gas temperature rise correlations and models with data from actual structures which were instrumented and burned in collaboration with the Kitchener Fire Department. According to the authors, the CFAST model was used 'successfully' to predict these conditions and will be used in future design of such facilities.

A report by Bailey et al. [116] compares predictions by CFAST version 3.1 to data from real scale fire tests conducted on board ex-USS SHADWELL, the Navy's R&D damage control platform. The phenomenon of particular interest in this validation series was the conduction of heat in the vertical direction through compartment ceilings and floors. As part of this work, Bailey et al. [117] compared CFAST temperature predictions on the unexposed walls of large metal

---

[2] Unless otherwise noted, percent differences are defined as (model-experiment)/experiment x 100.

boxes, driven by steady state fires. This tested the models prediction of radiation and conduction in both the vertical and horizontal directions. Indirectly it quantifies the quality of the conduction/convection/radiation models. The model and experiment compared well within measurement error bounds of each. The comparison was particularly good for measurements in the fire compartment as well as for the compartment and deck directly above it, with predictions typically agreeing with experiments within measurement uncertainty. The model under-predicted the temperatures of the compartments and decks not directly adjacent to the fire compartment early in the tests. Most of the error arose due to uncertainty in modeling the details of the experiment. The size of the vent openings between decks and to the outside must be included, but these were not always known. Cracks formed in the deck between the fire compartment and the compartment above due to the intense fire in the room of origin, but a time dependent record was not kept. The total size of the openings to the outside of warped doors in both compartments was not recorded. As can be seen in figures 7 and 8 of reference [116], the steady state predictions are identical (within error bounds of the experiment and prediction). The largest error is after ignition (uncertainty in the initial fire) and during development of the cracks between the compartments. While this does not affect the agreement in the room of origin, it does lead to an uncertainty of about 30% in the adjacent compartment.

6.2 Comparisons with Previously Published Test Data

A number of researchers have studied the level of agreement between computer fire models and real-scale fires. These comparisons fall into two broad categories: fire reconstruction and comparison with laboratory experiments. Both categories provide a level of verification for the models used. Fire reconstruction, although often more qualitative, provides a higher degree of confidence for the user when the models successfully simulate real-life conditions. Comparisons with laboratory experiments, however, can yield detailed comparisons that can point out weaknesses in the individual phenomena included in the models.

Deal[118] reviewed four computer fire models (CCFM[3], FIRST[119], FPETOOL[120] and FAST[2] version 18.5 (the immediate predecessor to CFAST)) to ascertain the relative performance of the models in simulating fire experiments in a small room (about 12 m^3 in volume) in which the vent and fuel effects were varied. Peak fire size in the experiments ranged up to 800 kW. According to the author, all the models simulated the experimental conditions including temperature, species generation, and vent flows 'quite satisfactorily.' With a variety of conditions, including narrow and normal vent widths, plastic and wood fuels, and flashover and sub-flashover fire temperatures, competence of the models at these room geometries was 'demonstrated.'

NIST has studied the predictive capability of CFAST in detail for several scenarios where experimental data were available. Peacock et al. [121] compared the performance of the CFAST model with experimental measurements for the variables presented above. Using a range of laboratory tests, they presented comparisons of peak values, average values, and overall curve shape for a number of variables of interest to model users. A total of five different real-scale fire tests were selected for the comparisons to represent a range of challenges for the CFAST model. Details of the experimental measurements and procedure for model calculations are available in the original paper [121]. Typical agreement between model predictions and experimental values ranged from about 2% to 25%. Careful specification of a simulation and building leakage were seen as important factors in assuring an accurate prediction.

6.2.1 NIST/NRC Verification and Validation

The U.S. Nuclear Regulatory Commission performed an extensive verification and validation of several fire models commonly used in nuclear power plant applications [9]. These models included simple spreadsheet calculations, zone models (including CFAST), and CFD models. The results of this study are presented in the form of relative differences between fire model predictions and experimental data for fire modeling attributes such as plume temperature that are important to NPP fire modeling applications. While the relative differences sometimes show agreement, they also show both under-prediction and over-prediction in some circumstances. These relative differences are affected by the capabilities of the models, the availability of accurate applicable experimental data, and the experimental uncertainty of these data. The two-zone models performed well when compared with the experiments considered. Evaluation of the two-zone models showed that the models simulated the experimental results within experimental uncertainty for most of the parameters of interest. The reason for this may be that the relatively simple experimental configurations selected for this study conform well to the simple two-layer assumption that is the basis of these models. However, users must remain cautious when applying these models to more complex scenarios, or when predicting certain phenomena, like heat fluxes. The results and comparisons included the the NRC study are included in the CFAST Software Development and Experimental Evaluation Guide for the current version of CFAST [8].

6.2.2 Fire Plumes

Davis compared predictions by CFAST version 5 (and other models) for high ceiling spaces [36]. In this paper, the predictive capability of two algorithms designed to calculate plume centerline temperature and maximum ceiling jet temperature in the presence of a hot upper layer were compared to measurements from experiments and to predictions using CFAST's ceiling jet algorithm. The experiments included ceiling heights of 0.58 m to 22 m and heat release rates of 0.62 kW to 33 MW. When compared to the experimental results CFAST's ceiling jet algorithm tended to over-predict the upper layer temperature by 20 %. With proper adjustment for radiation effects in the thermocouple measurements, some of this difference disappears. The effect of entrainment of the upper layer gases was identified for improvement.

CFAST includes the calculation of plume centerline temperature from Davis' work.

6.2.3 Multiple Compartments

Jones and Peacock [122] presented a limited set of comparisons between the FAST model (version 18.5) and a multi-room fire test. The experiment involved a constant fire of about 100 kW in a three-compartment configuration of about 100 m^3. They observed that the model predicted an upper layer temperature that was too high by about 20 % with satisfactory prediction of the layer interface position. These observations were made before the work of Pitts et al. [112] showed that the thermocouple measurements need to be corrected for radiation effects. Convective heating and plume entrainment were seen to limit the accuracy of the predictions. A comparison of predicted and measured pressures in the rooms showed within 20%. Since pressure is the driving force for flow between compartments, this agreement was seen as important.

Levine and Nelson [123] used a combination of full-scale fire testing and modeling to simulate a fire in a residence. The 1987 fire in a first-floor kitchen resulted in the deaths of three persons in an upstairs bedroom, one with a reported blood carboxyhemoglobin content of 91%. Considerable physical evidence remained. The fire was successfully simulated at full scale in a fully-instrumented seven-room two-story test structure. The data collected during the test have been used to test the predictive abilities of two multiroom computer fire models: FAST and HARVARD VI. A coherent ceiling layer flow occurred during the full-scale test and quickly carried high concentrations of carbon monoxide to remote compartments. Such flow is not directly accounted for in either computer code. However, both codes predicted the carbon monoxide buildup in the room most remote from the fire. Prediction of the pre-flashover temperature rise was also 'good' according to the authors. Prediction of temperatures after flashover that occurred in the room of fire origin was seen as 'less good.' Other predictions of conditions throughout the seven test rooms varied from 'good approximations' to 'significant deviations' from test data. Some of these deviations are believed to be due to combustion chemistry in the hot upper layer not considered in detail in either of the two models.

6.2.4 Large Compartments

Duong [124] studied the predictions of several computer fire models (CCFM, FAST, FIRST, and BRI [125]), comparing the models with one another and with large fires (4MW to 36MW) in an aircraft hanger (60000 m^3). For the 4MW fire size, he concluded that all the models are 'reasonably accurate.' At 36MW, however, 'none of the models did well.' Limitations of the heat conduction and plume entrainment algorithms were thought to account for some of the inaccuracies.

6.2.5 Prediction of Flashover

A chaotic event that can be predicted by mathematical modeling is that of flashover. Flashover is the common term used for the transition a fire makes from a few objects pyrolyzing to full room involvement. It is of interest to the fire service because of the danger to firefighters and to building designers because of life safety and the attendant impact on occupants. Several papers have looked at the capability of CFAST to predict the conditions under which flashover can occur.

Chow [126] concluded that FAST correctly predicted the onset of flashover if the appropriate criteria were used. The criteria were gas temperature near the ceiling, heat flux at the floor level and flames coming out of the openings. This analysis was based on a series of compartment fires.

A paper by Luo et al. [127] presents a comparison of the results from CFAST version 3 against a comprehensive set of data obtained from one flashover fire experiment. The experimental results were obtained from a full-scale prototype apartment building under flashover conditions. Three polyurethane mattresses were used as fuel. It was found that the predicted temperatures from the CFAST fire model agreed well with the experimental results in most areas, once radiation corrections are applied to the thermocouple data.

Collier [128] makes an attempt to quantify the fire hazards associated with a typical New Zealand dwelling with a series of experiments. These tests, done in a three-bedroom dwelling, included both non-flashover and flashover fires. The predictions by CFAST version 2 were seen by the authors as consistent with the experiments within the uncertainty of each.

Post-flashover fires in shipboard spaces have a pronounced effects on adjacent spaces due to highly conductive boundaries. CFAST (version 3.1) predictions for the gas temperature and the cold wall temperature were compared with shipboard fires [129]. The comparisons between the model and experimental data show 'conservative predictions' according to the authors. The authors attribute this to an overestimation of the average hot wall temperature and an underestimation of external convective losses due to wind effects.

Finally, a comparison of CFAST with a number of simple correlations was used by Peacock and Babrauskas [130, 131] to simulate a range of geometries and fire conditions to predict the development of the fire up to the point of flashover. The simulations represent a range of compartment sizes and ceiling heights. Both the correlations and CFAST predictions were seen to provide a lower bound to observed occurrence of flashover. For very small or very large compartment openings, the differences between the correlations, experimental data, and CFAST predictions was more pronounced.

The important test of all these prediction methods is in the comparison of the predictions with actual fire observations. Figure 6.1 (reference [131]) presents estimates of the energy required to achieve flashover for a range of room and vent sizes. This figure is an extension of the earlier work of Babrauskas [132] and includes additional experimental measurements from a variety of sources, most notably the work of Deal and Beyler [133]. For a number of the experimental observations, values are included that were not explicitly identied as being a minimum value at flashover. In addition, figure 6.1 includes predictions from the CFAST model (version 5).

As with some of the experimental data defining flashover as an upper layer temperature reaching 600 °C, many experimental measures were reported as peak values rather than minimum values necessary to achieve flashover. Thus, ideally all the predictions should provide a lower bound for the experimental data. Indeed, this is consistent with the graph the vast majority of the experimental observations lie above the correlations and model predictions. For a considerable range in the ratio $A_T/A\sqrt{h}$, the correlations of Babrauskas [132] Thomas [134], and the MQH correlation of McCaffrey et al. [135] provide similar estimates of the minimum energy required to produce flashover. The estimates of Hägglund [136] yields somewhat higher estimates for values of $A_T/A\sqrt{h}$ greater than 20 m$^{-1/2}$.

The results from the CFAST model for this single compartment scenario provides similar results to the experiments and the correlations for most of the range of $A_T/A\sqrt{h}$. For small values of $A_T/A\sqrt{h}$, the CFAST values rise somewhat above the values from the correlations. These small values of $A_T/A\sqrt{h}$ result from either very small compartments (small A_T) or very large openings (large $A_T/A\sqrt{h}$), both of which stretch the limits of the assumptions inherent in the model. For very small compartments, radiation from the fire to the compartment surfaces becomes more important, enhancing the conductive heat losses through the walls. However, the basic two-zone assumption may break down as the room becomes very small. For very large openings, the calculation of vent flow via an orifice flow coefficient approach is likely inaccurate. Indeed, for such openings, this limitation has been observed experimentally [132]. The estimates are close to the range of uncertainty shown by the correlations which also diverge at very small values of $A_T/A\sqrt{h}$.

Perhaps most significant in these comparisons is that all the simple correlations provide estimates similar to the CFAST model and all the models are consistent with a wide range of experimental data. For this simple scenario, little is gained with the use of the more complex models.

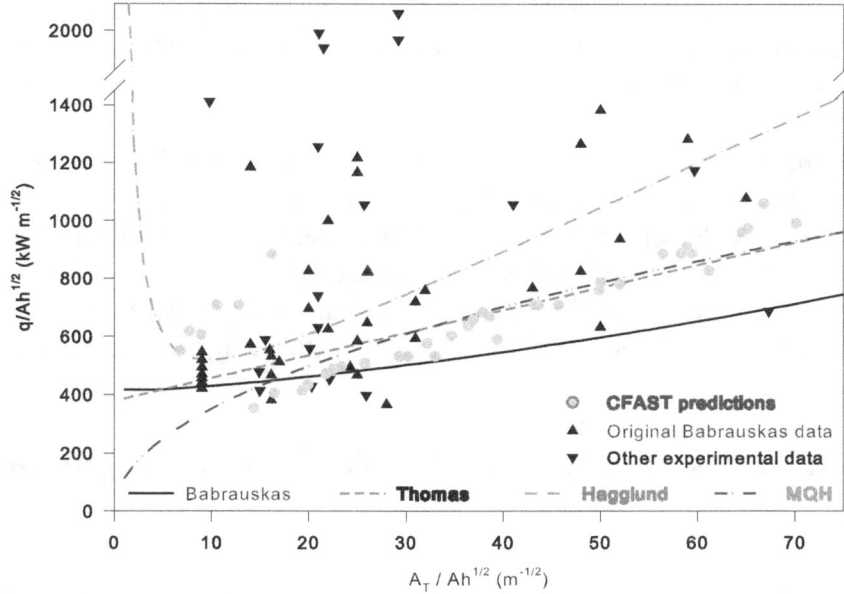

Figure 6.1: Comparison of correlations, CFAST predictions, and experimental data for the prediction of flashover in a compartment fire.

For more complicated scenarios, the comparison may not be as simple.

6.3 Comparison with Documented Fire Experience

There are numerous cases of CFAST being used to adjudicate legal disputes. Since these are discussed in courts of law, there is a great deal of scrutiny of the modeling, assumptions, and results. Most of these simulations and comparisons are not available in the public literature. A few of the cases which are available are discussed below. The metric for how well the model performed is its ability to reproduce the time-line as observed by witnesses and the death of occupants or the destruction of property as was used in evidence in legal proceedings.

As mentioned in section 6.2.3, Levine and Nelson describe the use of FAST for understanding the deaths of two adults in a residence in Sharon, Pennsylvania in 1987 [123]. The paper compared the evidence of the actual fire, a full scale mockup done at NIST and the results from FAST (version 18) [76] and Harvard VI [137]. The most notable shortcoming of the models was the lower than actual temperatures in the bedrooms, caused by loss of heat through the fire barriers. This led to the improvement in CFAST in the mid-90s to couple compartments together so that both horizontal and vertical heat transfer can occur to adjacent compartments.

Bukowski used CFAST version 3.1 to analyze a fire in New York City [138] in 1994 which resulted in the death of three firefighters. The CFAST model was able to reproduce the observed conditions and supported the theory as to how the fire began and the cause of death of the three fire fighters.

Chow describes the use and comparison of CFAST simulations with a 1996 high rise building fire in Hong Kong [139]. CFAST simulations were performed to help understand the probable fire environment under different conditions. Three simulations were performed to study the consequences of a fire starting in the lift shaft. Smoke flow in the simulations qualitatively matched those observed during the incident.

In the early morning hours of March 25, 1990 a tragic fire took the lives of 87 persons at a neighborhood club in the Bronx, New York [140]. The New York City Fire Department requested the assistance of the NIST Center for Fire Research (CFR) in understanding the factors which contributed to this high death toll and to develop a strategy that might reduce the risk of a similar occurrence in the many similar clubs operating in the city. The simulation showed the potential for development of untenable conditions within the club and particularly in the single exit stairway.

6.4 Comparison with Experiments Which Cover Special Situations

There are several sets of comparisons used in the development of the model or specific applications beyond those discussed more generally above.

6.4.1 Nuclear Facilities

Floyd validated CFAST version 3.1 by comparing the modeling results with measurements from fire tests at the Heiss-Dampf Reaktor (HDR) facility [141]. The structure was originally the containment building for a nuclear power reactor in Germany. The cylindrical structure was 20 m in diameter and 50 m in height topped by a hemispherical dome 10 m in radius. The building was divided into eight levels. The total volume of the building was approximately 11000 m^3. From 1984 to 1991, four fire test series were performed within the HDR facility. The T51 test series consisted of 11 propane gas tests and three wood crib tests. To avoid permanent damage to the test facility, a special set of test rooms were constructed, consisting of a fire room with a narrow door, a long corridor wrapping around the reactor vessel shield wall, and a curtained area centered beneath a maintenance hatch. The fire room walls were lined with firebrick. The doorway and corridor walls had the same construction as the test chamber. Six gas burners were mounted in the fire room. The fuel source was propane gas mixed with 10% air fed at a constant rate to one of the six burners.

In general, the comparison between CFAST and the HDR results was seen as 'good' by the author, with two exceptions. The first is the overestimate of the temperature of the upper layer, typically within about 15% of the experimental measurements. This is common and generally results from using too low a value for the production of soot, water (hydrogen) and carbon monoxide. The other exception consists of predictions in spaces where the zone model concept breaks down, for example in the stairways between levels. In this case, CFAST has to treat the space either in the filling mode (two layer approximation) or as a fully mixed zone (using the SHAFT option). Neither is quite correct, and in order to understand the condition in such spaces in detail (beyond the transfer of mass and energy), a more detailed CFD model must be used, for example,

FDS [47].

The U.S. Nuclear Regulatory Commission performed an extensive verification and validation of several fire models commonly used in nuclear power plant applications [142]. These models included simple spreadsheet calculations, zone models (including CFAST [9]), and CFD models. The results of this study are presented in the form of relative differences between fire model predictions and experimental data for fire modeling attributes such as temperature or heat flux that are important to NPP fire modeling applications. These relative differences are affected by the capabilities of the models, the availability of accurate applicable experimental data, and the experimental uncertainty of these data. Evaluation of the two-zone models showed that the models simulated the experimental results within experimental uncertainty for many of the parameters of interest. The reason for this may be that the relatively simple experimental configurations selected for this study conform well to the simple two-layer assumption that is the basis of these models.

While the relative differences sometimes show agreement for many parameters, they also show both under-prediction and over-prediction in some circumstances, most notably when conditions vary within a compartment or detailed local conditions are important to accurate prediction (for example, plume temperature or heat flux near to the fire source). The results and comparisons included the the NRC study are included in this report for the current version of CFAST.

6.4.2 Small Scale Testing

As an implementation of the zone model concept, CFAST is applicable to a wide range of scenarios. One end of this spectrum are small compartments, one to two meters on a side. Several research efforts have looked at small scale validation. There are three papers by Chow [143, 144, 145] which examine this issue. The first is the use of an electric heater with adjustable thermal power output was to verify temperature predictions by CFAST version 3.1. The second was closed chamber fires studied by burning four types of organic liquids, namely ethanol, N-heptane, and kerosene. The burning behavior of the liquids was observed, and the hot gas temperature measured. These behaviors along with the transient variations of the temperature were then compared with those predicted by the CFAST model. Finally, in another series of experiments, three zone models, one of which was CFAST, were evaluated experimentally using a small fire chamber. Once again, liquid fires were chosen for having better control on the mass loss rate. The results on the development of smoke layer and the hot gas temperature predicted by the three models were compared with those measured experimentally. According to Chow, 'fairly good agreement' was found if the input parameters were carefully chosen.

6.4.3 Unusual Geometry and Specific Algorithms

A zone model is inherently a volume calculation. There is an assumption in the derivation of the equations that gas layers are strongly stratified. This allows for the usual interpretation that a volume can then be thought of as a rectangular parallelepiped, which allows the developers to express the volume in terms of a floor area and height of a compartment, saying simply that the height times the floor area is the volume. However, there are other geometries which can be adequately described by zone models. Tunnels, ships, and attics are the most common areas of application which fall outside of the usual scope.

Railway and Vehicle Tunnels

Altinakar et al. [146] used a *modified version* of CFAST for predicting fire development and smoke propagation in vehicle or railroad tunnels. The two major modifications made to the model dealt with mixing between the upper and lower layers and friction losses along the tunnel. The model was tested by simulating several full-scale tests carried out at memorial Tunnel Ventilation Test Program in West Virginia, and the Offeneg Tunnel in Switzerland. His article compares simulated values of temperature, opacity and similar sensible quantities with measured values and discusses the limits of the applicability of zone models for simulating fire and smoke propagation in vehicle and railroad tunnels.

Peacock et al. [147] compared times to untenable conditions determined from tests in a passenger rail car with those predicted by CFAST for the same car geometry and fire scenarios. For a range of fire sizes and growth rates, they found agreement that averaged approximately 13%.

Non-Uniform Compartments

In January 1996, the U.S. Navy began testing how the CFAST model would perform when tasked with predicting shipboard fires. These conditions include mass transport through vertical vents (representing hatches and scuttles), energy transport via conduction through decks, improvement to the radiation transport sub-model, and geometry peculiar to combat ships. The purpose of this study was to identify CFAST limitations and develop methods for circumnavigating these problems [148]. A retired ship representing the forward half of a USS Los Angeles class submarine was used during this test. Compartments in combat ships are not square in floor area, nor do they have parallel sides.

Application of CFAST to these scenarios required a direct integration of compartment cross-sectional area as a function of height to correctly interpret the layer interface position and provide correct predictions for flow through doors and windows (vertical vents). This required user specification of the area as a function of height (ROOMA and ROOMH inputs) to provide a description for the model to use. For most applications of CFAST, the effort required for the input outweighs any additional precision in the calculated results gained by use of the ROOMA and ROOMH inputs in the model.

Long Corridors

Prior to development of the corridor flow model, the implementation of flow in compartments assumed that smoke traveled instantly from one side of a compartment to another. The work of Bailey et al. [149] provided the basis for the corridor flow model in CFAST. According to the author, it shows 'good agreement' for the delay time calculated using CFAST version 5 and measured flow along high aspect ratio passageways.

Mechanical Ventilation

There have been two papers which have looked at the effectiveness of the mechanical ventilation system. The first considered a fire chamber of length 4.0 m, width 3.0 m and height 2.8 m with adjustable ventilation rates [150]. Burning tests were carried out with wood cribs and methanol to study the preflashover stage of a compartmental fire and the effect of ventilation. The mass loss rate of fuel, temperature distribution of the compartment and the air intake rate were measured. The heat release rates of the fuel were calculated and the smoke temperature was used as a validation parameter. A scoring system was proposed to compare the results predicted by the three models. According to the author, CFAST does 'particularly well,' though there are some differences which can be attributed to the zone model approach.

A second series of experiments by Luo [151] indicate that the CFAST model (version 3.1) generally overpredicts the upper layer temperature in the burn room because the two-zone assumption is likely to break down in the burn room. It was found that the room averaged temperatures obtained from CFAST were in 'good overall agreement' with the experimental results. The discrepancies can be attributed to the correction needed for thermocouple measurements. The CO concentration, however, was inconsistent. CFAST tended to overestimate CO concentration when the air handling system was in operation. This was seen due to inconsistencies in what is measured (point measurements) and predicted (global measurements).

Sprinkler Activation

A suppression algorithm [66] was incorporated into CFAST. Chow [152] evaluates the predictive capability for a sprinkler installed in an atrium roof. There were three main points being considered: the possibility of activating the sprinkler, thermal response, and water requirement. The zone model CFAST was used to analyze the possibility of activation of a sprinkler head. Results derived from CFAST were seen to be 'accurate, that is, providing good agreement with experimental measurements.'

t^2 Fires

Matsuyama conducted a series of full-scale experiments [153] using t^2 fires. Fire room and corridor smoke filling processes were measured. The size of the corridors and arrangements of smoke curtains were varied in several patterns. Comparisons were then made between the experimental results and those predicted by CFAST. The author concludes that while the model does a 'good job' of predicting experimental results, there are systematic differences which could be reduced with some revision to zone model formulation to include the impact of smoke curtains.

6.5 Summary

How to best quantify the comparisons between model predictions and experiments is not obvious. The necessary and perceived level of agreement for any variable is dependent upon both the typical use of the variable in a given simulation, the nature of the experiment, and the context of the comparison in relation to other comparisons being made. For instance, the user may be interested in the time it takes to reach a certain temperature in the room, but have little or no interest in peak temperature for experiments that quickly reach a steady-state value. Insufficient experimental data and understanding of how to compare the numerous variables in a complex fire model prevent a complete validation of the model.

A true validation of a model would involve proper statistical treatment of all the inputs and outputs of the model with appropriate experimental data to allow comparisons over the full range of the model. Thus, the comparisons of the differences between model predictions and experimental data discussed here are intentionally simple and vary from test to test and from variable to variable due to the changing nature of the tests and typical use of different variables. Table 6.1 summarizes the Validation comparisons included for the current version of the model detailed in the Software Development and Experimental Evaluation Guide for CFAST [8].

Table 6.1: Summary of Model Comparisons

Quantity	Average Difference[a] (%)	Median Difference[b] (%)	Within Experimental Uncertainty[c] (%)	90th Percentile[d] (%)
HGL Temperature	6	14	52	30
HGL Depth	3	15	40	28
Plume Temperature	17	11	39	29
Ceiling Jet Temperature	16	5	70	61
Oxygen Concentration	-6	18	12	32
Carbon Dioxide Concentration	-16	16	21	52
Smoke Obscuration[e]	272/22	227/18	0/82	499/40
Pressure	43	13	77	206[f]
Target Flux (Total)	-23	27	42	51
Target Temperature	0	18	38	34
Surface Flux (Total)	5	25	40	61
Surface Temperature	24	35	17	76

a - averaged difference includes both the sign and magnitude of the relative differences in order to show any general trend to over- or under-prediction.
b - median difference is based only on the magnitude of the relative differences and ignores the sign of the relative differences so that values with opposing signs do not cancel and make the comparison appear closer than individual magnitudes would indicate.
c - the percentage of model predictions that are within experimental uncertainty.
d - 90 % of the model predictions are within the stated percentage of experimental values. For reference, a difference of 100 % is a factor of 2 larger or smaller than experimental values.
e - the first number is for the closed door NIST/NRC tests and the second number if for the open door NIST/NRC tests.
f - high magnitude of the 90th percentile value driven in large part by two tests where under-prediction was approximately 2 Pa.

CFAST predictions in this validation study were consistent with numerous earlier studies, which show that the use of the model is appropriate in a range of fire scenarios. The CFAST model has been subjected to extensive evaluation studies by NIST and others. Although differences between the model and the experiments were evident in these studies, most differences can be explained by limitations of the model as well as of the experiments. Like all predictive models, the best predictions come with a clear understanding of the limitations of the model and the inputs provided to perform the calculations.

Chapter 7
Conclusion

CFAST is a collection of data, computer programs, and documentation which are used to simulate the important time-dependent phenomena describing the character of a compartment fire. The major functions provided include calculation of the buoyancy-driven as well as forced transport of energy and mass through a series of specified compartments and connections (e.g., doors, windows, cracks, ducts), and the resulting temperatures, smoke optical densities, and gas concentrations after accounting for heat transfer to surfaces and dilution by mixing with clean air.

CFAST is a zone model. The basic assumption of all zone fire models is that each compartment can be divided into a small number of control volumes, each of which is internally uniform in temperature and composition. Beyond these basic assumptions, the model typically involves a mixture of established theory (e.g., conservation equations), empirical correlations where there are data but no theory (e.g., flow and entrainment coefficients), and approximations where there are neither (e.g., post-flashover combustion chemistry) or where their effect is considered secondary compared to the "cost" of inclusion (e.g., temperature dependent material properties)..

The predictive equations are based on the fundamental laws of conservation of mass and energy. Empirical correlations are employed to bridge gaps in existing knowledge. Since the necessary approximations required by operational practicality result in the introduction of uncertainties in the results, the user should understand the inherent assumptions and limitations of the programs, and use these programs judiciously - including sensitivity analyses for the ranges of values for key parameters in order to make estimates of these uncertainties.

As discussed in this report, the CFAST model has been subjected to extensive evaluation studies by NIST and others. Although differences between the model and the experiments were evident in these studies, most differences can be explained by limitations of the model as well as of the experiments. Like all predictive models, the best predictions come with a clear understanding of the limitations of the model and of the inputs provided to do the calculations.

CFAST has proven to be fast, robust and reliable. While the focus of the development of the model has been whole building simulations for assessing the effect of fire on a building environment, principally to calculate threats to life safety of occupants and insults to the building structure, it has been used for a wide variety of building and fire scenarios. The simplest use has been to ascertain the sufficiency of an air handling system to extract smoke. The most complex has been an assessment of fire propagation in a high-rise complex. It is also widely used as the fire model in egress calculations and is described as the basis for hazard estimates in the Simulex[154] and Exodus[155] egress models.

Because of the speed of the model, it is possible to do real parameter studies of the building environment. It is reasonable to do actual parameter studies including the tens of thousands of variations needed for a proper hazard and risk calculation. Even in those cases where more detailed predictions are needed (e.g., smoke detector and sprinkler head siting), CFAST provides the capability to scope the problem, in essence doing parameter studies to determine what specific scenario should be addressed by more detailed calculations.

References

[1] American Society for Testing and Materials, West Conshohocken, Pennsylvania. *ASTM E 1355-04, Standard Guide for Evaluating the Predictive Capabilities of Deterministic Fire Models*, 2004.

[2] W. W. Jones and R. D. Peacock. Technical Reference Guide for FAST Version 18. Technical Note 1262, National Institute of Standards and Technology, 1989.

[3] L. Y. Cooper and G. P. Forney. The Consolidated Compartment Fire Model (CCFM) Computer Application CCFM-VENTS – Part I: Physical Reference Guide. NISTIR 4342, National Institute of Standards and Technology, 1990.

[4] R. W. Bukowski, R. D. Peacock, W. W. Jones, and C. L. Forney. Technical Reference Guide for the HAZARD I Fire Hazard Assessment Method. Version 1.1. Volume 2. NIST Handbook 146/II, National Institute of Standards and Technology, 1991.

[5] R. D. Peacock, W. W. Jones, G. P. Forney, R. W. Portier, P. A. Reneke, R. W. Bukowski, and J. H. Klote. Update Guide for HAZARD I Version 1.2. NISTIR 5410, National Institute of Standards and Technology, 1994.

[6] W. W. Jones, R. D. Peacock, G. P. Forney, and P. A. Reneke. CFAST – Consolidated Model of Fire Growth and Smoke Transport (Version 6): Technical Reference Guide. Special Publication 1026, National Institute of Standards and Technology, Gaithersburg, Maryland, July 2008.

[7] R. D. Peacock, W. W. Jones, P. A. Reneke, and G. P. Forney. CFAST – Consolidated Model of Fire Growth and Smoke Transport (Version 6): User's Guide. Special Publication 1041, National Institute of Standards and Technology, Gaithersburg, Maryland, December 2005.

[8] R. D. Peacock, K. B. McGrattan, B. Klein, W. W. Jones, and P. A. Reneke. CFAST – Consolidated Model of Fire Growth and Smoke Transport (Version 6): Software Development and Model Evaluation Guide. Special Publication 1086, National Institute of Standards and Technology, Gaithersburg, Maryland, November 2008.

[9] Verification and Validation of Selected Fire Models for Nuclear Power Plant Applications, Volume 5: Consolidated Fire and Smoke Transport Model (CFAST),. NUREG 1824, U.S. Nuclear Regulatory Commission, Office of Nuclear Regulatory Research, Rockville, MD, 2007.

[10] F. P. Incorpera and D. P. DeWitt. *Fundamentals of Heat Transfer*. John Wiley and Sons, 1981.

[11] G. P. Forney and W. F. Moss. Analyzing and Exploiting Numerical Characteristics of Zone Fire Models. *Fire Science and Technology*, 14(1 and 2):49–60, 1994.

[12] R. G. Rehm and G. P. Forney. A Note on the Pressure Equations Used in Zone Fire Modeling. NISTIR 4906, National Institute of Standards and Technology, 1992.

[13] K. D Steckler, J. G. Quintiere, and W. J. Rinkinen. Flow Induced by Fire in a Compartment. NBSIR 82-2520, National Bureau of Standards, 1982.

[14] J. G. Quintiere, K. DSteckler, and D. Corley. An Assessment of Fire Induced Flows in Compartments. *Fire Science and Technology*, 4(1), 1984.

[15] J. H. Klote. Fire Experiments of Zoned Smoke Control at the Plaza Hotel in Washington, DC. NISTIR 90-4253, National Institute of Standards and Technology, 1990.

[16] W. W. Jones and R. D. Peacock. Using CFAST to Estimate the Efficiency of Filtering Particulates in a Building. NISTIR 7498, National Institute of Standards and Technology, 2008.

[17] J. H. Morehart, E. E. Zukoski, and T. Kubota. Characteristics of Large Diffusion Flames Burning in a Vitiated Atmosphere. In *Third International Symposium on Fire Safety Science*, Edinburgh, 1991.

[18] W. M. Thornton. The Relation of Oxygen to the Heat of Combustion of Organic Compounds. *Philosophical Magazine and Journal of Science*, 33(6th Series):196–203, 1917.

[19] C. Huggett. Estimation of the Rate of Heat Release by Means of Oxygen Consumption Calorimetry. *Journal of Fire and Flammability*, 12:61, 1980.

[20] G. Heskestad. Fire Plumes, Flame Height, and Air Entrainment. In *SFPE Handbook of Fire Protection Engineering*, 3rd Ed. National Fire Protection Association and The Society of Fire Protection Engineers, 2002.

[21] R. A. Bryant, T. J. Ohlemiller, E. L. Johnsson, A. Hamins, B. S. Grove, F. Guthrie, A. Maranghides, and G. W. Mullholland. NIST 3 Megawatt Quantitative Heat Releae Rate Facility. Special Publication 1007, National Institute of Standards and Technology, 2003.

[22] V. Babrauskas, J. R. Lawson, W. D. Walton, and W. H. Twilley. Upholstered Furniture Heat Release Rates Measured with a Furniture Calorimeter. NISTIR 82-2604, National Institute of Standards and Technology, 1982.

[23] V. Babrauskas and J. F. Krasny. Fire Behavior of Upholstered Furniture. Monograph 173, National Bureau of Standards, 1985.

[24] B. T. Lee. Effect of Ventilation on the Rates of Heat, Smoke, and Carbon Monoxide Production in a Typical Jail Cell Fire. NBSIR 82-2469, National Bureau of Standards, 1982.

[25] Z. Alterman. Effect of Surface Tension on the Kelvin-Helmholtz Instability of Two Rotating Fluids. *Proceedings of the National Academy of Sciences (USA)*, 47(2):224–227, 1961.

[26] D. Drysdale. *An Introduction to Fire Dynamics*. John Wiley and Sons, New York, 1985.

[27] A. Tewarson. Combustion of Methanol in a Horizontal Pool Configuration. Technical Report RC78-TP-55, Factory Mutual Research Corporation, Norwood, MA, 1978.

[28] B. J. McCaffrey. Entrainment and Heat Flux of Buoyant Diffusion Flames. NBSIR 82-2473, National Bureau of Standards, 1982.

[29] H. Koseki. Combustion Properties of Large Liquid Pool Fires. *Fire Technology*, 25:241, 1989.

[30] B. R. Morton, G. Taylor, and J. S. Turner. Turbulent Gravitational Convection from Maintained and Instantaneous Sources. *Proceedings of the Royal Society of London. Series A, Mathematical and Physical Sciences*, 234(1196):1–23, January 24 1956.

[31] B. J. McCaffrey. Momentum Implications for Buoyant Diffusion Flames. *Combustion and Flame*, 52:149, 1983.

[32] G. Heskestad. Engineering Relations for Fire Plumes. *Fire Safety Journal*, 7:25–32, 1984.

[33] B. M. Cetegen, E. E. Zukoski, and T. Kubota. Entrainment and Flame Geometry of Fire Plumes. NBS-GCR 82-402, National Bureau of Standards, 1982.

[34] B. M. Cetegen, E. E. Zukoski, and T. Kubota. Entrainment in the Near and Far Field of Fire Plumes. *Combustion Science and Technology*, 39(1-6):305–331, 1984.

[35] E. E. Zukoski, T. Kubota, and B. M. Cetegen. Entrainment in Fire Plumes. *Fire Safety Journal*, 3:107–121, 1981.

[36] W. D. Davis. Comparison of Algorithms to Calculate Plume Centerline Temperature and Ceiling Jet Temperature With Experiments. *Journal of Fire Protection Engineering*, 12:9, 2002.

[37] H. R. Baum and B. J. McCaffrey. Fire Induced Flow Field: Theory and Experiment. In T. Wakamatsu, Y. Hasemi, A. Sekizawa, and P. G. Seeger, editors, *Fire Safety Science. Proceedings. 2nd International Symposium*, pages 129–148, Tokyo, Japan, June 13-17 1989. International Association for Fire Safety Science, Hemisphere Publishing Corporation.

[38] D. D. Evans. Calculating Fire Plume Characteristics in a Two Layer Environment. *Fire Technology*, 20(3):39–63, 1984.

[39] R. D. Peacock, S. Davis, and B. T. Lee. Experimental Data Set for the Accuracy Assessment of Room Fire Models. NBSIR 88-3752, National Bureau of Standards, 1988.

[40] E. E. Zukoski, T. Kubota, and C. S. Lim. Experimental Study of Environment and Heat Transfer in a Room Fire. Mixing in Doorway Flows and Entrainment in Fire Plumes. NBS-GCR 85-493, National Bureau of Standards, 1985.

[41] J. G. Quintiere, K. D Steckler, and B. J. McCaffrey. A Model to Predict the Conditions in a Room Subject to Crib Fires. In *First Specialist Meeting (International) of the Combustion Institute*, Talance, France, 1981.

[42] L. Y. Cooper. Calculation of the Flow Through a Horizontal Ceiling/Floor Vent. NISTIR 89-4052, National Institute of Standards and Technology, 1989.

[43] L. Y. Cooper. Algorithm and Associated Computer Subroutine for Calculating Flow Through a Horizontal Ceiling/Floor Vent in a Zone-Type Compartment Fire Model. NISTIR 4402, National Institute of Standards and Technology, 1990.

[44] L. Y. Cooper. Combined Buoyancy- and Pressure-Driven Flow Through a Shallow, Horizontal Circular VENT. *Journal of Heat Transfer*, 117:659–667, August 1995.

[45] J. H. Klote and J. A. Milke. *Principles of Smoke Management*. American Society of Heating, Refrigerating, and Air-Conditioning Engineers, Inc, Atlanta, GA, 2002.

[46] *2001 ASHRAE Handbook - HVAC Systems and Equipment*. American Society of Heating, Refrigerating, and Air-Conditioning Engineers, Inc, Altanta, GA, 2001.

[47] K. B. McGrattan, S. Hostikka, J. E. Floyd, H. R. Baum, and R. G. Rehm. Fire Dynamics Simulator (Version 5), Technical Reference Guide. NIST Special Publication 1018-5, National Institute of Standards and Technology, Gaithersburg, Maryland, October 2007.

[48] G. P. Forney. Computing Radiative Heat Transfer Occurring in a Zone Fire Model. NISTIR 4709, National Institute of Standards and Technology, 1991.

[49] R. Siegel and J. R. Howell. *Thermal Radiation Heat Transfer*. Hemisphere Publishing Corporation, New York, 2nd edition, 1981.

[50] H. C. Hottel. *Heat Transmission*. McGraw-Hill Book Company, New York, 3rd edition, 1954.

[51] H. C. Hottel and E. Cohen. Radiant Heat Exchange in a Gas Filled Enclosure: Allowance for Non-uniformity of Gas Temperature. *American Institute of Chemical Engineering Journal*, 4(3), 1958.

[52] T. Yamada and L. Y. Cooper. Algorithms for Calculating Radiative Heat Exchange Between the Surfaces of an Enclosure, the Smoke Layers and a Fire. July 1990.

[53] W. W. Jones and G. P. Forney. "Improvement in Predicting Smoke Movement in Compartmented Structures. *Fire Safety Journal*, 21:269, 1993.

[54] C. L. Tien, K. Y. Lee, and A. J. Stretton. Radiation Heat Transfer. In P. J. DiNenno, editor, *SFPE Handbook of Fire Protection Engineering*, chapter 1-4. National Fire Protection Association and The Society of Fire Protection Engineers, 3rd edition, 2002.

[55] C. L. Tien and G. Hubbard. Infrared Mean Absorption Coefficients of Luminous Flames and Smoke. *Journal of Heat Transfer*, 100:235, 1978.

[56] D. K. Edwards. Radiation Properties of Gases. In W. M. Rohsenow, editor, *Handbook of Heat Transfer Fundementals*, pages 74–75. McGraw-Hill Book Company, 2nd edition, 1985.

[57] H. C. Hottel. Radiant Heat Transmission. In W. H. McAdams, editor, *Heat Transmission*. McGraw-Hill Book Company, New York, 1942.

[58] A. Atreya. Convection Hear Transfer. In P. J. DiNenno, D. Drysdale, C. L. Beyler, and W. D. Walton, editors, *SFPE Handbook of Fire Protection Engineering*, chapter 1-3. National Fire Protection Association and The Society of Fire Protection Engineers, Quincy, MA, 3rd edition, 2003.

[59] G. H. Golub and J. M. Ortega. *Scientific Computer and Differential Equations. An Introduction to Numerical Methods*. Academic Press, New York, 1989.

[60] W. F. Moss and G. P. Forney. "Implicitly coupling heat conduction into a zone fire model. NISTIR 4886, National Institute of Standards and Technology, 1992.

[61] L. Y. Cooper. Fire-Plume-Generated Ceiling Jet Characteristics and Convective Heat Transfer to Ceiling and Wall Surfaces in a Two-Layer Zone-Type Fire Environment. NISTIR 4705, National Institute of Standards and Technology, 1991.

[62] L. Y. Cooper. Heat Transfer in Compartment Fires Near Regions of Ceiling-Jet Impingement on a Wall. *Journal of Heat Transfer*, 111:455, 1990.

[63] L. Y. Cooper. Ceiling Jet-Driven Wall Flows in Compartment Fires. *Combustion Science and Technology*, 62:285, 1988.

[64] Y. Jaluria and L. Y. Cooper. "Negatively Buoyant Wall Flows Generated in Enclosure Fires. *Progress in Energy and Combustion Science*, 15:159, 1989.

[65] G. Heskestad and H. F. Smith. Investigation of a New Sprinkler Sensitivity Approval Test: the Plunge Test. Technical Report 22485 2937, Factory Mutual Research Corporation, Norwood, MA, 1976.

[66] D. Madrzykowski and R. Vettori. A Sprinkler Fire Suppression Algorithm for the GSA Engineering Fire Assessment System. NISTIR 4883, National Institute of Standards and Technology, 1992.

[67] D. D. Evans. Sprinkler Fire Suppression for HAZARD. NISTIR 5254, National Institute of Standards and Technology, 1993.

[68] D. Madrzykowski. Evaluation of Sprinkler Activation Prediction Methods. In *International Conference on Fire Science and Engineering*, Kowlon, 1995. ASIAFLAM, Interscience Limited.

[69] J. Seader and I. Einhorn. "Some Physical, Chemical, Toxicological and Physiological Aspects of Fire Smokes. In *Sixteenth Symposium (International) on Combustion*, pages 1423–1445, Pittsburgh, PA, 1976. The Combustion Institute.

[70] G. W. Mullholland and C. Croarkin. Specific Extinction Coefficient of Flame Generated Smoke. *Fire and Materials*, 24:227, 2000.

[71] F. M. Galloway and M. M. Hirschler. A Model for the Spontaneous Removal of Airborne Hydrogen Chloride by Common Surfaces. *Fire Safety Journal*, 14:251, 1989.

[72] F. M. Galloway and M. M. Hirschler. "Transport and Decay of Hydrogen Chloride: Use of a Model to Predict Hydrogen Chloride Concentrations in Fires Involving a Room-Corridor-Room Arrangement. *Fire Safety Journal*, 16:33, 1990.

[73] W. W. Jones, G. P. Forney, R. D. Peacock, and P. A. Reneke. A Technical Reference Guide for CFAST: An Engineering Tool for Estimating Fire and Smoke Transport. Technical Note 1431, National Institute of Standards and Technology, 2003.

[74] R. D. Peacock, G. P. Forney, P. A. Reneke, R. W. Portier, and W. W. Jones. CFAST, The Consolidated Model of Fire Growth and Smoke Transport. Technical Note 1299, National Institute of Standards and Technology, 1993.

[75] W. W. Jones. Modeling Smoke Movement Through Compartmented Structures. *Journal of Fire Sciences*, 11(2):172, 1993.

[76] W. W. Jones. Multicompartment Model for the Spread of Fire, Smoke and Toxic Gases. *Fire Safety Journal*, 9(1):172, 1985.

[77] W. W. Jones. "Prediction of Corridor Smoke Filling by Zone Models. *Combustion Science and Technology*, 35:229, 1984.

[78] W. W. Jones and G. P. Forney. Modeling Smoke Movement Through Compartmented Structures. In *Proceedings of the Fall Technical Meeting of the Combustion Institute, Eastern States Section*, Ithaca, NY, 1991.

[79] R. D. Peacock, G. P. Forney, and P. A. Reneke. Improved Numerics and Structure in CFAST. Unpublished Memorandum, February 1992.

[80] R. D. Peacock. Fix for Chemistry Algorithm in CFAST. Unpublished Memorandum, April 11 1994.

[81] R. D. Peacock. New Convection Algorithm in CFAST. Unpublished Memorandum, January 5 1993.

[82] R. D. Peacock. HCl Deposition. Unpublished Memorandum, January 5 1993.

[83] R. D. Peacock. New Output and History File Formats for CFAST. Unpublished Memorandum, September 9 1993.

[84] W. W. Jones. Internal memo dated February 1996 Announcing Version 3.0 of CFAST. February 1996.

[85] W. W. Jones. Internal memo dated November 1997 Announcing Version 3.1.1 of CFAST. November 1997.

[86] W. W. Jones. Differences in the 4.0.1 Code from the Beta Release of September 1, 1999 to March 1, 2000. March 2000.

[87] Fire Hazard User's Group. Description of the CFAST 1.4 Release. Letter to CFAST Users, December 1991.

[88] Fire Hazard User's Group. Description of the CFAST 1.6 Release. Letter to CFAST Users, October 1992.

[89] Fire Hazard User's Group. Description of the CFAST 2.0 Release. Letter to CFAST Users, October 1993.

[90] Fire Hazard User's Group. Description of the CFAST 2.0.1 Release. Letter to CFAST Users, January 1996.

[91] Fire Hazard User's Group. Description of the CFAST 3.0 Release. Letter to CFAST Users, August 1996.

[92] W. D. Walton. Zone Computer Fire Models for Enclosures. In P. J. DiNenno, D. Drysdale, C. L. Beyler, and W. D. Walton, editors, *SFPE Handbook of Fire Protection Engineering*, chapter 3-7. National Fire Protection Association and The Society of Fire Protection Engineers, Quincy, MA, 3rd edition, 2003.

[93] *NFPA 805, Performance-Based Standard for Fire Protection for Light Water Reactor Electric Generating Plants*. 2004/2005 National Fire Codes. National Fire Protection Association, Quincy, MA, 2001 edition, 2004.

[94] *NFPA 551, Guide for the Evaluation of Fire Risk Assessment*. 2004/2005 National Fire Codes. National Fire Protection Association, Quincy, MA, 2004 edition, 2004.

[95] J. R. Barnett and C. L. Beyler. "Development of an Instructional Program for Practicing Engineers HAZARD I Users. NBS-GCR 90-580, National Institute of Standards and Technology, 1990.

[96] K. E. Brenan, S. L. Campbell, and L. R. Petzold. *Numerical Solution of Initial-Value Problems in Differential-Algebraic Equations*. Elsevier Science Publishing, New York, 1989.

[97] R. L. Iman and J. C. Helton. An Investigation of Uncertainty and Sensitivity Analysis Techniques for Computer Models. *Rick Analysis*, 8(1):71–90, 1988.

[98] R. D. Peacock, P. A. Reneke, C. L. Forney, and M. M. Kostreva. Issues in Evaluation of Complex Fire Models. *Fire Safety Journal*, 30:103–136, 1988.

[99] A. Beard. Evaluation of Fire Models: Part I – Introduction. *Fire Safety Journal*, 19:295–306, 1992.

[100] K. A. Notarianni. *The Role of Uncertainty in Improving Fire Protection Regulation*. PhD thesis, Carnegie Mellon University, Pittsburgh, PA, 2000.

[101] N. Khoudja. *Procedures for Quantitative Sensitivity and Performance Validation of a Deterministic Fire Safety Model*. PhD thesis, Texas A&M University, 1988.

[102] G.E.P. Box, W.G. Hunter, and J.S. Hunter. *Statistics for Experimenters, An Introduction to Design, Data Analysis and Model Building*. John Wiley and Sons, New York, 1978.

[103] C. Daniel. *Applications of Statistics to Industrial Experimentation*. John Wiley and Sons, New York, 1976.

[104] A.M. Walker. Uncertainty Analysis of Zone Fire Models. Research Report 97/8, University of Canterbury, New Zealand, 1997.

[105] H.W. Emmons. Why Fire Model? *Fire Safety Journal*, 13:77, 1988.

[106] D.Q. Duong. The Accuracy of Computer Fire Models: Some Comparisons with Experimental Data from Australia. *Fire Safety Journal*, 16:415, 1990.

[107] Verification and Validation of Selected Fire Models for Nuclear Power Plant Applications, Volume 1: Main Report. NUREG 1824, U.S. Nuclear Regulatory Commission, Office of Nuclear Regulatory Research, Rockville, MD, 2007.

[108] *NFPA 72, National Fire Alarm Code*. 2004/2005 National Fire Codes. National Fire Protection Association, Quincy, MA, 2003 edition, 2003.

[109] V. Babrauskas and R.D. Peacock. Heat Release Rate: The Single Most Important Variable in Fire Hazard. *Fire Safety Journal*, 18:255, 1992.

[110] N.A. Dempsey, P.J. Pagni, and R.B. Williamson. Compartment Fire Experiments: Comparison With Models. *Fire Safety Journal*, 25(3):187, 1995.

[111] International Conference of Building Officials, Whittier, CA. *1997 Uniform Building Code*, 1997.

[112] W.M. Pitts, E. Braun, R.D. Peacock, H.E. Mitler, E.L. Johnsson, P.A. Reneke, and L.G. Blevins. Temperature Uncertainties for Bare-Bead and Aspirated Thermocouple Measurements in Fire Environments. In *Thermal Measurements: The Foundation of Fire Standards*, Special Technical Publication 1427, West Conshohocken, PA, 2001. American Society for Testing and Materials.

[113] Y. He and V. Beck. Smoke Spread Experiment in a Multi-Storey Building and Computer Modeling. *Fire Safety Journal*, 28(2):139, 1997.

[114] M. Luo, Y. He, and V. Beck. Comparison of Existing Fire Model Predictions With Experimental Results From Real Fire Scenarios. *Journal of Applied Fire Science*, 6(4):357, 1996/1997.

[115] G. M. Poole, E. J. Weckman, and A. B. Strong. Fire Growth Rates in Structural Fires. NISTIR 5499, National Institute of Standards and Technology, 1994.

[116] J. Bailey and P. Tatem. Validation of Fire/Smoke Spread Model (CFAST) Using Ex-USS SHADWELL Internal Ship Conflagration Control (ISCC) Fire Tests. Technical Report NRL/MR/6180-95-7781, Naval Research Laboratory, 1995.

[117] J. L. Bailey, P. A. Tatem, W. W. Jones, and G. P. Forney. Development of an Algorithm to Predict Vertical Heat Transfer Trough Ceiling/Floor Conduction. *Fire Technology*, 34(2):139, 1998.

[118] S. Deal. A Review of Four Compartment Fires with Four Compartment Fire Models. In *Proceedings of the Annual Meeting of the Fire Retardant Chemicals Association*, volume Fire Safety Developments and Testing, pages 33–51, Ponte Verde Beach, Florida, October 21-24 1990. Fire Retardant Chemicals Association.

[119] H. E. Mitler and J. A. Rockett. Users' Guide to FIRST, A Comprehensive Single-Room Fire Model. NBSIR 87-3595, National Institute of Standards and Technology, September 1987.

[120] H. E. Nelson. FPETOOL: Fire Protection Engineering Tools for Hazard Estimation. NISTIR 4380, National Institute of Standards and Technology, 1990.

[121] R. D. Peacock, W. W. Jones, and R. W. Bukowski. Verification of a Model of Fire and Smoke Transport. *Fire Safety Journal*, 21:89–129, 1993.

[122] W. W. Jones and R. D. Peacock. Refinement and Experimental Verification of a Model for Fire Growth and Smoke Transport. In T. Wakamatsu, Y. Hasemi, A. Sekizawa, P. G. Seeger, P. J. Pagni, and C. E. Grant, editors, *Fire Safety Science. Proceedings. 2nd International Symposium*, pages 897–906, Tokyo, Japan, June 13-17 1988. International Association for Fire Safety Science, Hemisphere Publishing Corporation.

[123] R. S. Levine and H. E. Nelson. Full Scale Simulation of a Fatal Fire and Comparison of Results with Two Multiroom Models. NISTIR 90-4168, National Institute of Standards and Technology, 1990.

[124] D. Q. Duong. The Accuracy of Computer Fire Models: Some Comparisons with Experimental Data from Australia. *Fire Safety Journal*, 16:415, 1990.

[125] T. Tanaka. Model of Multiroom Fire Spread. *Fire Science and Technology*, 3(2):105–121, 1983.

[126] W. Chow. Predicability of Flashover by Zone Models. *Journal of Fire Sciences*, 16:335, September/October 1988.

[127] M. Luo, Y. He, and V. Beck. Application of Field Model and Two-Zone Model to Flashover Fires in a Full-Scale Multi-Room Single Level Building. *Fire Safety Journal*, 29:1, 1997.

[128] P. Collier. Fire in a Residential Building: Comparisons Between Experimental Data and a Fire Zone Model. *Fire Technology*, 32:195, August/September 1996.

[129] D. White, C. Beyler, J. Scheffrey, and F. Williams. Modeling the Impact of Post-Flashover Shipboard Fires on Adjacent Spaces. *Journal of Fire Protection Engineering*, 10, 2000.

[130] R. D. Peacock, P. A. Reneke, R. W. Bukowski, and V. Babrauskas. Defining Flashover for Fire Hazard Calculations. *Fire Safety Journal*, 32(4):331–345, 1999.

[131] V. Babrauskas, R. D. Peacock, and P. A. Reneke. Defining Flashover for Fire Hazard Calculations: Part II. *Fire Safety Journal*, 38:613–622, 2003.

[132] V. Babrauskas. Upholstered Furniture Room Fires–Measurements, Comparison with Furniture Calorimeter Data, and Flashover Predictions. *Journal of Fire Sciences*, 2(1):5–19, January/February 1984.

[133] S. Deal and C. Beyler. Correlating Preflashover Room Fire Temperatures. *Journal of Fire Protection Engineering*, 2(2):33–48, 1990.

[134] P. H. Thomas. Testing Products and Materials for Their Contribution to Flashover in Rooms. *Fire and Materials*, 5:103–111, 1981.

[135] B. J. McCaffrey, J. G. Quintiere, and M. F. Harkleroad. Estimating Room Temperatures and the Likelihood of Flashover Using Fire Test Data Correlations. *Fire Technology*, 17(2):98–119, 1981.

[136] B. Hägglund. Estimating Flashover Potential in Residential Rooms. FOA rapport C 202369-A3, Forsvarets Forkningsanstalt, 1980.

[137] J. A. Rockett and M. Morita. The NBS Harvard VI Multi-room Fire Simulation. *Fire Science and Technology*, 5(2):159–164, 1985.

[138] R. W. Bukowski. Modeling a Backdraft Incident: The 62 Watts Street (New York) Fire. *Fire Engineers Journal*, 56(185):14–17, 1996.

[139] W. Chow. Preliminary Studies of a Large Fire in Hong Kong. *Journal of Applied Fire Science*, 6(3):243–268, 1996/1997.

[140] R. W. Bukowski. Analysis of the Happyland Social Club Fire with HAZARD I. *Fire and Arson Investigator*, 42:36, 1992.

[141] J. Floyd. Comparison of CFAST and FDS for Fire Simulation With the HDR T51 and T52 Tests. NISTIR 6866, National Institute of Standards and Technology, 2002.

[142] Verification and Validation of Selected Fire Models for Nuclear Power Plant Applications, Volume 1: Main Report. NUREG 1824, U.S. Nuclear Regulatory Commission, Office of Nuclear Regulatory Research, Rockville, MD, 2007.

[143] G. Lui and W. Chow. A Short Note on Experimental Verification of Zone Models with an Electric Heater. *International Journal on Engineering Performance-Based Fire Codes*, 5:30, 2003.

[144] W. Chow. Studies on Closed Chamber Fires. *Journal of Fire Sciences*, 13:89, 1995.

[145] W. Chow. Experimental Evaluation of the Zone Models CFAST, FAST and CCFM.VENTS. *Journal of Applied Fire Science*, 2:307, 1992-1993.

[146] M. Altinakar, A. Weatherhill, and P. Nasch. Use of a Zone Model in Predicting Fire and Smoke Propagation in Tunnels. In J. R. Gillard, editor, *9th International Symposium on Aerodynamics and Ventilation of Vehicle Tunnels*, pages 623–639, Aosta Valley, Italy, October 6-8 1997. BHR Group.

[147] R. D. Peacock, J. D. Averill, D. Madrzykowski, D. W. Stroup, P. A. Reneke, and R. W. Bukowski. Fire Safety of Passenger Trains; Phase III: Evaluation of Fire Hazard Analysis Using Full-Scale Passenger Rail Car Tests. NISTIR 6563, National Institute of Standards and Technology, 2004.

[148] J. Hoover and P. A. Tatem. Application of CFAST to Shipboard Fire Modeling. Part 3. Guidelines for Users. NRL/ML 6180-01-8550, Naval Research Laboratory, 2001.

[149] J. L. Bailey, G. P. Forney, P. A. Tatem, and W. W. Jones. Development and Validation of Corridor Flow Submodel for CFAST. *Journal of Fire Protection Engineering*, 12:139, 2002.

[150] W. Chow. Use of Zone Models on Simulating Compartmental Fires With Forced Ventilation. *Fire and Materials*, 14:466, 1995.

[151] M. Luo. One Zone or Two Zones in the Room of Fire Origin During Fires? *Journal of Fire Sciences*, 15:240, 1997.

[152] W. Chow. Performance of Sprinkler in Atria. *Journal of Fire Sciences*, 14:466, 1996.

[153] K. Matsuyama, T. Wakamatsu, and K. Harada. Systematic Experiments of Room and Corridor Smoke Filling for Use in Calibration of Zone and CFD Fire Models. NISTIR 6588, National Institute of Standards and Technology, 2000.

[154] P. A. Thompson, J. Wu, and E. W. Marchant. Modelling Evacuation in Multi-storey Buildings with Simulex. *Fire Engineering*, 56(185):7–11, 1996.

[155] S. Gwynne, E. R. Galea, P. Lawrence, and L. Filippidis. Modelling Occupant Interaction with Fire Conditions Using the building EXODUS Evacuation Model. Technical Report 00/IM/54, University of Greenwich, 2000.

www.ingramcontent.com/pod-product-compliance
Lightning Source LLC
Chambersburg PA
CBHW081727170526
45167CB00009B/3734